TALES OF TW

A doctor's life in Liverpool and Hong Kong

By

Chris Howard

This book is dedicated to all those who taught me so much, all of them: my parents, my teachers, my family, my patients and my students.

And to Kate who was the catalyst.

"The blessed will not care what angle they are regarded from, having nothing to hide."

– W.H.Auden.

"Gentleness is not always kindness."

– Penelope Fitzgerald.

ABOUT THE AUTHOR

Chris Howard is a long-term Hong Kong resident. He trained as a doctor in Liverpool in the 1960s and in 1974 joined a family medical practice in Hong Kong.

A keen yachtsman and a mediocre skier he is also an existential threat to his wife Joan's garden plants. His three daughters and eight grandchildren are scattered around the Globe so he travels extensively. He splits his time between Hong Kong, Devon and SW France.

This is his first book.

TABLE OF CONTENTS

BOOK 1. STUDENT DAYS.

CHAPTER 1. AN AEROGRAM.

Grab a chance and you won't be sorry for a might have been.   Arthur Ransome.

Looking back, I can see that the turning point of my life was on that cold afternoon when I stood on the Headrow in the Yorkshire city of Leeds and having a few minutes to kill I pulled up my collar against the cold. I reached into an overcoat pocket to take out my copy of the BMJ that had arrived that morning in the post. Just as always, I opened it to the back pages and idly searched down the columns of small ads for 'Positions Vacant: Abroad' while I stood in the queue waiting for a bus back to the hospital. A cruel wind blew from the moors and snow was forecast for later.

One entry caught my eye at once: a doctor in Hong Kong wanted to recruit a paediatrician to join his family practice. In my imagination I could suddenly feel the tropical heat on my shoulders and sense the bright colours and dark mysterious shadows, the noise and the smells of the Far East. Like one of Conrad's heroes, I would stride through crowded streets hearing strange languages, smelling the tree blossoms and the aromas of food cooking in roadside stalls... and the drains. This small ad seemed to be calling out to me. I knew it was a life I wanted and I was suddenly determined to seize my chance to live it.

I abandoned the queue and with my mind awhirl with thoughts of adventure, went into the Post Office and borrowed a pen at the counter to write a flimsy blue aerogramme and send it off to this Bill Oram at his clinic in Hong Kong.

It was 1974 and I was a medical registrar at the General Infirmary in Leeds, working in the paediatric department. A very good job it was too, with plenty to learn, good juniors, well trained nursing staff and a youngish boss who was keen to teach. Its downside was the long hours of overtime with little time off and also responsibilities split among three separate hospitals spread across the city.

1

Until that winter I had lived and worked in Liverpool for ten years: five as a student and then a further five years after qualification as a junior doctor training in Liverpool's hospitals. I was not very happy about moving away from my adopted home town and all my friends, but across the Pennines in Yorkshire the new job was a good opportunity. I was content enough in this new city, the clinical work was varied and interesting and I was thinking of how I might fit in some research. So, I buzzed from hospital to hospital in my MGB sportscar, with only the occasional feeling of loneliness in a strange new place while I waited for Joan and our daughter Diane to arrive and move into the small town-house at St. James's Hospital. Joan had found a job in the lab at the children's heart surgery unit nearby.

Britain was entering a deep recession following the Middle East oil crisis and the recent fall of the Conservative government, broken by a year of unmanageable strikes and belligerent trade unionism. Disgracefully, there had even been a doctors' strike. Harold Wilson's new Labour government intended to cut public spending and as that had already started with the NHS further job promotion in England seemed unlikely for a while and finding the academic or research post I needed next, impossible. My professor advised me to look abroad for a university posting, Africa perhaps or the USA. There seemed to be plenty of opportunities in Africa but funding was being cut back there too and my enquiries were dealt with at a snail's pace.

I called Joan who I expected, even hoped, would pour cold water on my plans; plans that even to me were beginning to appear ridiculous. What did I know about Asia or Hong Kong, not having even travelled outside Europe? I had no idea what medical practice was like there or whether I would be able to do some research. Another thought nagged at me that Hong Kong was said to be fabulously expensive where everyone lived in tiny apartments crowded onto hillsides? I knew nothing, I had never even met anyone who had been there. And what about schools for Diane? However, I did know that I liked both warm weather and Chinese cuisine. Joan however turned out to be cautiously interested and suggested that we took a wait-and-see attitude, reminding me that perhaps I would not be offered the job at all.

2

So, I got on with my work. We all enjoyed the little house at Jimmy's[1] which was close by the neonatal unit. I joined the local sailing club and fetched my boat from Liverpool. I had begun to forget about the delights of the Far East when a blue aerogram arrived, postmarked Hong Kong: 'Please send us a photo. When can you start? Terms of employment to follow' and so on. Of course, I agreed to everything and planned to leave Leeds at the end of June. It was a leap in the dark and looking back I am amazed at Joan and myself; we just sold up, said goodbye to everyone and, with the clothes on our backs, plus a few books and a washer-dryer to follow later by sea, left our old lives. The famous washer-dryer eventually arrived after great expense in our new home. It promptly broke down and was scrapped.

The heat and humidity were a great shock, not at all the balmy warmth I had imagined while I was still in Leeds. The streets were not just thronged; I had never experienced such a hustle and bustle as on the crowded pavements in Central. The smells were as I anticipated: of cooking, of drains, of diesel exhausts, of tree blossom. We rented a temporary flat near Stanley on the south side of the island overlooking a sandy beach. We planned to stay there for a few months while we searched for an apartment that we could afford on a salary that was much smaller than it had seemed in England. Hong Kong soon lived up to its reputation as an expensive place to live.

My new partner was a New Zealander, a surgeon of wide experience who had also been trained in obstetrics. On my second day in Hong Kong, I went to meet him at the clinic in Central. It was situated in an office block on Des Voeux Road, one floor above a busy Chinese bank. Bill Oram was a big man in more ways than one. He had a powerful personality of the type that when he arrived in a room everyone seemed to turn towards him. He and I were to work together for many years, not always smoothly but always with mutual respect. He was certainly pleased to see me, having had to work single-handed for several months since my predecessor Peter had abruptly left and returned to England for reasons never fully explained.

---

[1] St. James's Teaching Hospital was known to one and all as 'Jimmy's'.

Bill immediately disapproved of my loose fitting 'tropical' suit of light linen of which I was so proud. Bought especially for life in the Far East from a Leeds tailor, it was apparently quite wrong for professional life locally. In retrospect I think I must have looked a bit like Somerset Maugham, only lacking the straw hat. Everyone either wore a proper business suit and tie or a 'safari suit'. As that was what Bill Oram wore in the summer season, so should I. A few days later I took the Star Ferry across the Harbour to the famous Sam the Tailor on Nathan Road to be measured for my safari suits while Joan looked through swatches of worsted cloth all the way from Huddersfield. Sam only needed a couple of days to have me fitted out and at such a low price that I began to think that life here was not going to be as expensive as we feared.

Joan with Bill Oram.

On that first day, Bill showed me how the clinic worked and introduced me to Vivienne, our receptionist. She was tall for a Chinese woman and later I was to recognise that hers were the typical striking facial features of a northerner, probably with Manchu ancestry; although her family had lived in Guangzhou before moving to Hong Kong when Vivienne was still a youngster. She proved a cheerful and intelligent woman on whom I was going to rely for many years to come. She showed me the appointment book for the next day, which was already full. Then Bill loaded me into his Jaguar and we roared off through the traffic on a rapid tour of Hong Kong; zipping through the appropriately named mid-levels half way up the slope between Central and the Peak; we visited the three hospitals Bill liked to use, and on to the south side of Hong Kong Island looked in on the practice's second clinic in Repulse Bay. We drove through the residential areas where the expatriates preferred to live and stopped at both the Yacht Club and the Aberdeen Boat Club which suggested it wasn't to be all work and no leisure. A cup of tea at his mansion of a flat in Kotewall Road in the Mid-Levels completed the tour and I was sent off to the Lee Gardens in Causeway Bay to get myself registered on the medical practitioners' list.

Lee Gardens proved a horticultural disappointment with not a blade of grass or a leaf, let alone an exotic flower, to be seen among its crowded streets. At the Health Department's office, I showed my passport and my medical diplomas from England; I signed some forms and parted with a small fee. They gave me my 'chop': a rubber stamp[2] with my registration number and that was it. I was a Hong Kong doctor.

Next day I would start work, but that evening I drove back to our leave-flat in Bill's old Mini Cooper with the windows open, as small cars did not usually boast air-conditioners then. I cooled off with a swim from the nearby beach with Joan and Diane. Diane loved to splash in the water but hated the sand sticking to her skin, Chung Hom Kok beach was very sandy! That evening Joan and I tried to relax, sweating under the ceiling fans and wondering what our new life would bring in this strange and enticing Chinese city.

I had come a long way from the nineteen-year-old applying to medical school a decade before, perhaps I should go back to that day in 1962.

---

[2] Chop. A rubber stamper. Essential for all Hong Kong business affairs. No document is complete without at least one stamp.

CHAPTER 2. YOUTH.

Liverpool; the Mersey and the three 'Graces' at Pier Head.

I could smell the River Mersey on that grey morning as I walked up the hill from Lime Street Station towards the medical school. A brisk west wind whipped through town and waste paper blew across my path to be caught, flapping on a rusty wire fence on the other side of the road. The pavement was cracked and stained, the tarmac on the street potholed. On my right, ringed with rusty chain-fencing was a big abandoned site; its buildings had been demolished by Nazi bombs 20 years before and it still lay empty. To the left stood a 'bullring'-shaped tenement, its walls were defaced by graffiti and its pavements littered with broken glass and rubbish. Though 1962 saw the height of the Liverpool-Sound, the Cavern and the Beatles, the city itself was a sorry place. The docks, previously its lifeblood, were worn down with strikes and lack of investment. The unions had opposed the planned container port which was now being built in Holland ready to take away what remained of Merseyside's shipping business. The great shipyards in Birkenhead across the river were idle; Liverpool was famous for its strikers; old businesses were closing or moving to the South-East of the country.

The once proud city looked into an uncertain future of poverty and unemployment. I could see this clearly and I'd only been there an hour. What was I thinking? Did I really want to move from the leafy Cheshire countryside to this grim city to spend 5 years or so on the toughest course any university could offer? Did I? Yes, indeed I did.

Ahead, at the top of the rise I could see the gleaming chocolate-brown tiles of the University's Victoria Building which dated from the times when the city was still brimming with confidence and money. Successful merchants had been proud of their achievements and wanted to show the world. Liverpool was then the great mercantile gate into Britain, and into Europe too. The entrance for cargos and passengers crossing the Atlantic Ocean, not just from the Americas but from everywhere on the globe. The era that built the 'Three Graces', those proud buildings standing side by side at the Pier Head, was only a few decades past but somehow, since the end of the war, Liverpool had lost its way, it had refused to change and had gone into decline.

I had to walk past the Victoria Building and turn left in front of the library and then past a yard filled with heaps of coal for the Royal Infirmary's boilers; and there was the Medical School. At the time it was considered an impressive piece of modern architecture with a lecture theatre dramatically cantilevered out from one side of the building. I walked up the stairs to the first floor and was greeted by a sign pointing to the interview room. A smiling secretary gave me a cup of tea and on a row of hard chairs I sat, uncomfortable in my ill-fitting suit, wearing one of my dad's ties, waiting nervously until I was called into the dragons' den.

Doctor this, professor that, the Dean and a severe looking woman dressed in a man's tweed suit; I was introduced to each one in turn. All the questions that I had expected; where had I come from today, what books had I read recently, why did I want to be a doctor? They probed around my unexpectedly poor performance in the A-level chemistry exam and, since I was spending my gap-year working as an apprentice pharmacist, why wouldn't I prefer the pharmacology course? A few more informal questions and 'We'll be in touch'.

Back at home I alternated between hope and despair as I recalled my answers. However, the committee had seemed pleased with what I had said, they did seem friendly; even the scary tweedy lady who turned out to be a gynaecologist. I didn't have to wait long for their letter; it said that I could start in September and there was no need to re-sit my A-levels.

Abandoned, the Austin 7 and MG-PA sat in my parents' drive.

Many people seem to know from a young age that they want to be doctors, but I had not. Engineering had been my aspiration, to design and build engines and machines, or perhaps bridges and dams. I had renovated a couple of pre-war cars, first an Austin 7 and later an MG Midget. These old bangers now stood, leaking oil on my parents' driveway, abandoned. The fun had been in the repairs but I preferred to actually drive my mother's new red Mini.

However, engineering proved a hopeless ambition; my maths teacher and the headmaster persuaded me to forget it as my maths was just too weak; they urged me to change to biology and perhaps follow my parents into pharmacy. Food technology was another idea, but reading about this and after applying for a sandwich course somewhere in the industrial wastes of NE England, I became so depressed with that idea that I even considered emigrating to Australia for £5 on the S.S. Canberra. This all changed for me when on one Saturday afternoon in Macclesfield I found myself with some fellow Kings School sixth-form boys helping to run the annual Careers Conference for the girls at the nearby High-School. We had set out chairs and tables, arranged water-glasses, pinned up signs, brewed tea and then we hung around hoping to chat-up some of the girls. When the doors opened it seemed that most headed straight for the nursing stand, the Civil Service recruiting desk was busy, even the armed-forces stand was attracting attention. But one desk had no customers; at it sat a doctor from Manchester University looking a bit forlorn beside his posters showing a caduceus, the inevitable group of happy students in gowns and a stethoscope rampant. Feeling a bit sorry for him and at a loose end myself, I brought him a cup of coffee and was invited to sit down for a chat.

This was the turning point in my life! I had never even considered becoming a medic before, but within half an hour I now knew both that it was possible and that it was exactly what I had been searching after. Whyever had I not considered it before?

My father was not so easily persuaded; with three sons to educate he could imagine the escalating costs, not to mention the no small matter of a five-year course instead of the usual three. Unsurprisingly, he was not convinced about how serious I was about yet another sudden turn in my ambitions. I stuck to my guns and eventually he agreed; I expect I have in fact to thank my mother for changing his mind.

On reflection, at this distance in time, I think that another major influence on my choice was the pharmacy apprenticeship I had started when I was eighteen. I spent the year after leaving school working at Boots the Chemists' shop in Bramhall, a small dormitory town a few miles from home by local train towards Manchester. My job was mainly in the backroom; helping make up prescriptions, managing the stock and ordering from the warehouse. I kept the dispensary tidy and the dispenser supplied with tea and biscuits. In those days, pharmacies actually made up their own bottles of medicine, making mixtures from scratch or adding extra ingredients to stock medicines which arrived from the warehouse in Winchester size bottles. A doctor might prescribe that so much phenobarbitone or perhaps belladonna tincture be added to, say, Mist. Magnesium Trisilicate[3]. Then we would weigh out the additive and shake it into a 20-ounce bottle of the medicine. Sometimes the additive was only a colouring, since a pink placebo obviously works better than a white one! There was a pill rolling machine in the back of a cupboard but I never saw it in action; the dispenser reminisced about days gone by when doctors ordered special tablets, sometimes with silver or gold-leaf coatings for their well-off private patients.

Looking back, I realise that a lot of the drugs we handed out in the 1960s were pretty ineffective. For example, none of the drugs used today for indigestion or duodenal ulcers were available then; the poor patients suffered with only various chalky alkaline mixtures and bland diets. When their symptoms got too bad, or when the ulcer burst, they underwent one of a number of operations which only occasionally made things better. After that, the long-suffering victim would keep quiet about his symptoms for fear of worse interventions. It is almost a rule in medicine that where there is a number of alternative treatments or operations then probably none of them is of much use, if a treatment really works everyone plumps for just that one and the rest are consigned to history.

---

[3] Mist. Mag. Trisil. A white chalky medicine prescribed for indigestion.

Heart drugs like Digitalis and blood-pressure tablets such as Aldomet caused many side effects; the much reviled 'Big-Pharma' has made such a difference to all our lives and we have forgotten about the rather primitive treatments in use 60 years ago.

I was the dogsbody who wrote the labels and counted out the tablets to put into screw-top bottles; blister-packs were unknown and pills and capsules came in 1000 tablet jars, their bright colours on the shelves resembling a candy-store. At night after a day's counting, my dreams were of a whirl with multicoloured pills on the little triangular tray that I would use for counting, flicking them around with a special spatula.

When the dispensary was quiet, I served in the shop alongside the sales-girls who were all a little older than me and very much more worldly. Boots refused to sell contraceptives in those days on moral grounds and so whenever a man entered the shop looking embarrassed in the particular way they did then, the girls would push me forward to deal with him. I'd have to explain that Jesse Boot, our great founder in Nottingham, long dead, was opposed to sexual intercourse and then I would point him to the other chemist's shop down the street where with a front window dominated by the traditional three huge glass flasks full of red, amber and green liquid they didn't have such a silly attitude. Our front window, brightly lit, exhibited cosmetics and beauty products to the passing world; very modern. Every few weeks a young man came from head office to 'dress' this window.

Mr East the shop-manager would often give us pep-talks and tips on marketing. He was very keen on follow-up selling. For example, if a customer bought an Instamatic camera, we should press them to buy a film or flash-bulbs. He went round the group inviting suggestions along that line, the girls suggested offering mascara to someone buying lipstick and so on. When it was my turn, I offered a roll of toilet paper to a customer looking for cascara laxatives; this successfully broke up the meeting.

One of the dispensers had recently completed his National Service in the army and was still in the reserves. In October the Russians and the Americans seemed to be on the brink of war during the Cuban Missile Crisis. Everyone was genuinely worried about the possibility of nuclear confrontation and it was a nervous time as the young Jack Kennedy outfaced the older and wilier Nikita Kruschev. Our dispenser received a call-up letter ordering him to return to uniform and was in such a nervous state that he was unable to concentrate on his work and made several mistakes; eventually he agreed to take a few days off. As history relates, the Russians backed down and Armageddon was postponed yet again.

It was easy to make a mistake: drug names were often similar, minute amounts of powders had to be weighed on a special balance, patients' names got mixed up and of course the doctors' handwriting didn't help. One of our local GPs was a particular problem, he always wrote with a fountain-pen in green ink which was often smeared since he didn't seem to bother much with blotting paper. Frequently I had to phone his surgery for a translation. This chore always fell to me as this Dr G. had a fiery temper especially late in the day after a long clinic.

This job certainly helped me grow up. I was now in a world of adults and high standards were expected of me. I got to know the doctors and nurses in the Bramhall area and most importantly I learned how to deal with the general public, some friendly, others aggressive and difficult. This all helped to convince me that I really did want to work in healthcare and help people individually. Whatever my inevitable early doubts, I knew now that medicine was the right choice.

Dinghy sailing in North Wales.

That year quickly passed and at the end of July I left the job at Boots to enjoy a last few summer weeks of holiday before starting at university. I wanted to go dinghy sailing and had a happy fortnight racing in the Menai Straits regattas and then I went down to Buckinghamshire as a sailing instructor at the National Sports Centre on the Thames at Marlow. This was my first experience of any sort of teaching. Rain or shine, we sailed every day in Enterprise dinghies on the river in lovely surroundings, best of all I got paid for it.

CHAPTER 3. MEDICAL SCHOOL.

The anatomy they learn is sheer unscientific nonsense, but they still have to learn it. **C.P. Snow**

In Britain nowadays around 40% of school-leavers go on to university; in 1962 it was well under 10%. There were no student loans and instead the government gave cash grants to students worth about £350 annually, that is between £7 and £8 thousand in today's money. They paid the tuition fees as well. There was a catch however; the grants were means-tested according to the family income. As the scion of a moderately well-to-do family, I only got the £50 minimum grant and Dad had to make up the rest. After buying textbooks and a half skeleton, then paying for accommodation, meals and other essentials like travelling there was not much left from £350. 'Poorer' families happily added more pocket money, but my dad thought he had paid out enough so it was a good thing I had managed to save some money from my gap-year wages.

The small intake into tertiary education meant that almost everyone there was really quite bright. Most of us had been in the top flight of pupils in the sixth form but now we were just average, if that. Medicine, although the course lasted for 5 years, required concentrated effort from the very start; if you fell behind there was no hope of catching up. The schoolboy's habit of learning everything in the last few weeks of pre-exam revision would not work, now there was to be only one opportunity to learn the subject well. On one day's knowledge the course continually built the next layer. My new friends, perhaps doing Arts courses for example, could party and miss lectures knowing they could catch up with the reading later. Logically I knew this was my lot, but it was hard to accept particularly after I had taken a year away from education and had rather lost the habit of studying hard.

I arrived in Liverpool on a glorious early autumn afternoon. My mother had driven me from home in the red Mini. We stopped in Woolworth's to buy the necessaries to make coffee and tea, I really wanted a gramophone but that was an ask too far. I found my room in the Rathbone Hall of Residence and Mum drove off and left me to it. I pinned up a sailing poster on the wall, arranged my few books on the shelf, put my clothes in the wardrobe and then went exploring.

Rathbone was near the famous Penny Lane of the Beatles song, in a part of the rather smart Allerton village, a very different Liverpool from the city centre and the deprived area near the medical school. Several green parks surrounded us including the huge Sefton Park with its palm house and lakes. At the end of our street was Smithdown Road with bus stops, a large pub and a chippy. This last was to be an important source of late-night nutrition; it fuelled much burning of midnight oil as I swotted for tests and exams over the years.

As more students arrived in taxis from the station, others driven by parents or struggling on foot with their suitcases from the bus stop I went out to meet my fellows. In those days few students owned a motor car and I don't think my old MG would have made it as far as Liverpool anyway.

The first few days were to be a 'Freshers' Conference'. A crash-introduction to student life. We discovered the canteens and the bars in the Union building. The debating society put on a show, the student president, one Rodney Ledward,[4] later to become notorious for medical negligence on a large scale, made an amusing speech. Then we were encouraged to join a club. Stalls competed for our subscriptions; Gilbert and Sullivan singers, Bridge club, sports clubs, Asian student society, Christians of several colours. I went straight to the sailing club booth where they were actually building a racing dinghy as they tried to recruit new members.

On the second morning we medical students were corralled into the antique wooden stalls of the surgical lecture theatre in the Derby Building. This part of the medical school was built onto the back of the Royal Infirmary (LRI) and had an illustrious history: in its laboratories Sherrington discovered the synapse and Gregory elucidated the properties of the hormone Gastrin. 120 of us sat on the wooden benches in semi-circular

---

[4] Rodney Ledward FRCS, Gynaecologist accused of botching operations on 13 private patients and struck-off in 1998. He escaped worse punishment by dying soon after.

tiers looking down on the lecturer's rostrum. After a fine harangue about ethics, hard work and personal hygiene we were released in a happy horde, already making friendships which would last many decades.

Clutching a reading list and accompanied by my new friend Bill Freeston we headed for the used-book-store. As anatomy never changed much, if at all, we could all buy the old tomes hocked a few weeks earlier by the students in the year above us. Broken spines, jotted notes in the margins, underlining, splashes of adipose tissue from the dissection room and a few missing pages would be no bother for us. Gray's *Anatomy,* Zuckerman's *Guide to Human Dissection*, Le Gros-Clark's *Tissues of the Body,* Ham's *Histology*; it all weighed a ton. Then to a scruffy shop selling medical instruments, no, not for a stethoscope (that would be next year), but for a half-skeleton. A battered wooden box contained loose human bones; a skull, an articulated spine with the vertebrae all wired neatly together, a collection of ribs of various sizes, bones from arms, legs and half a pelvic bone, scapula and clavicle, a hand and a foot, both articulated. They had to be from the same person or they wouldn't match together. Who was she? For mine was indeed a young woman's skeleton. Probably very old, these bones must have helped many many students before me. I took her back to Rathbone; she was mine for the next three terms.

Bill was also living in Rathbone and we caught the bus back to the hall together with our bags of books and boxes of bones. We had a cup of tea in his room and then I went back to mine to start ploughing through the books. Next day was a big one, we were to be introduced to our cadavers.

I had heard of people 'leaving their bodies to science' and now I was to see what happened to some of them. Nowadays we are used to the idea of organ donation; the useful body parts: kidneys, liver, heart, corneas, are removed while the donor is still warm within minutes of death, then the donor's body can be released to the family and the funeral can be held without extra delays. Bodies given to anatomy schools, though, may remain unburied for months or years, they are carefully preserved by embalming with formalin and other chemicals and kept until needed for dissection.

In clean white coats, we all trooped into the anatomy theatre, a large space well-lit by high windows on both sides. Along the sides of the room stood twenty or more steel tables each with a plastic sheet covering a rounded lump. Directed to our tables in groups of eight in alphabetical order, Bill and I were again together. Four of us to each side of our body. The cadaver was anonymous, a number was inked on one foot so someone knew who he had been. He was well built and muscular, about 60 or 70 years old at a guess, from his big rough hands and the line of suntan at his collar we assumed he had been an outdoor manual worker, a farmer perhaps or a stevedore.

None of that really mattered to us, he had given us his earthly remains and we were to do our best to learn from them. I was more impressed by this notion than the fact of this being the first dead person I had encountered. I have read narratives by Richard Gordon and Patrick Taylor, for example, of their first experience in the anatomy rooms but I felt none of the apprehension or uneasiness that they described. In fact, the only negative feeling was from the preservatives that made our noses and eyes itch. I felt that there was a hell of a lot of work in front of us and I needed to get started.

We were to begin in the armpit, the axilla. This area is quite complicated comprising the shoulder joint with its complex articulations and the nerves that emerge from the neck and course into the arm through an intricate plexus. Muscles both moved and stabilised the shoulder joint; we had to know all this and to dissect away skin and fat to demonstrate all the features that are shown in the textbooks. We had to take it in turns and it took many hours of painstaking work. One of us, Graham, took the attitude that this was all a complete waste of time and so hardly did anything to assist, preferring to stay in his digs and study the books; so the rest of us had more to do. Perhaps the amount of detail we had to learn and the amount of drudgery doing the dissection really was poor use of our time. Many medical schools today no longer teach anatomy all in one year but bit by bit throughout the course as necessary. Computer models are replacing cadavers. Using MRI images students study anatomy on a screen and can highlight nerves or blood vessels, twist and turn the images for different views. Time marches on and we do things differently, occasionally better.

Once a week came the dreaded viva-voce. An anatomy demonstrator, usually a surgeon in training; or a lecturer; or worse still, the Professor, would come and sit by our body and grill us. We had to survive this inquisition before proceeding to the next stage of cutting. Vivas were usually on Friday afternoons and marked the end of the week. Friday night was our night off when we headed for the pub.

Not all the medics were in Hall, in fact most had accommodation in digs and were spread all over the city, some were still living at home, perhaps a train-ride away on the Wirral. These people probably only met students on the medical course, while those like Bill and I in Hall got to know students from the other faculties. On my corridor were a lawyer, a post-grad from Addis Ababa researching malaria, two mathematicians, an electrical engineer, an architect and a social scientist. I think we had more of a university experience as a result. Most of the non-medical students had much more free time, they usually lay in late each morning and were always ready for a chat or to go partying. They tended to disrupt my studying; it is very hard to concentrate for the nth time on the course of a nerve or the way a shoulder joint works when your friends are gathering noisily just outside your door on their way out for a few beers.

The result of this deprivation, this monkish devotion to study, was that when we did hit the town everyone knew it. I will spare you all the details! We usually started at a pub on a cul-de-sac near the medical school and just opposite the steps leading down to the Infirmary's VD clinic. It served the usual slightly warm and vinegary Walker's bitter and, fresh from passing our vivas, or sometimes failing them, we were ready to down a good dose of it. The landlord was celebrated for being able to drink many pints without taking a breath. We had all seen his X-rays showing large congenital pharyngeal pouches; hollow outgrowths from his throat extending down inside the chest and well able to contain the large volume of Walker's best that he chugged back. Shortly after showing off this feat, he then had to go and empty these pouches in the back toilets. Several years later, after we had left medical school, we heard that he developed a cancer in one of them and died in the LRI.

Student Rag-Day. Bill is on my right. The R.C. Cathedral site behind us.

The problem with Friday celebration was that we still had to be back in the lecture theatre on Saturday morning quite early and endure a few hours of lectures after which the weekend was our own. From Rathbone we went on walks in Sefton Park or down to Otterspool Promenade, an open area landscaped over old landfill alongside the Mersey. The brown muddy water swirled by, smelly and polluted from factories upstream in Widnes, but the views were lovely and we could see the Welsh hills in the distance; gleaming and clean.

On Sunday I liked to go sailing in the University's Fireflies on the Marine Lake at Southport. It was a long way by train and bus but I soon found someone with a car who gave me a lift most weeks if I paid for the petrol. I joined in the racing and also did my bit by giving lessons to beginners. The club had recently bought some new Firefly racing dinghies; the grant from the University was for 6 boats but by buying part-completed kits we could afford 8 or 9 of them. We took over a large storeroom in the Catholic Cathedral which was under construction next to the Union building which we could use as a workshop. At that time after decades of work, only Lutyens' magnificent crypt had been finished. In our nice warm space, the boats were decked, varnished and fitted out. The one I helped with was called 'Epolenep' after someone's girlfriend. When they were all launched 'Myfanwy' was my favourite, painted yellow, its woodwork was beautifully finished by a Dylan Thomas fan from the English Department who claimed the right to name her.

It was cold for sailing late in the year but it became even colder in the spring term and in February we had to break the ice at the edge of the lake to launch the boats. That winter is remembered as the 'great freeze', one of the coldest on record. Nowadays we sail in all weathers, wet-suits and dry-suits keep us warm and dry even if we capsize. In 1963 we just got wet and shivered watching our fingers go white then purple.

University Fireflies racing on a winter day at West Kirby.

Bill Brockbank[5],the captain of our first team, was one of the star sailors in the Firefly class, he was also something of a wild man. He felt there was a fault in the boat's design that good boat-handling technique should overcome; a Firefly will inevitably capsize if you try to gybe[6] it when planing[7] along at high speed. On a windy day when approaching the gybe mark in a race one is usually planing like mad with no choice in the matter, and you just have to gybe and hope for the best. One very draughty afternoon I was crewing for Bill and he wanted to try an idea he'd conceived for solving this problem. As we screamed towards the buoy, the hull half out of the water, he told me his plan, which I have now forgotten; but for my part I was to sit still, not moving at all and at a critical moment pull the centreboard right up. A few seconds later I was in the water under the sail, Bill was up on the top side of the overturned boat, of course perfectly dry. We bailed her out and tried again, three times, I think. I was almost unconscious from hypothermia when he agreed to give up. Believe me, a Firefly should not be gybed when going at high speed.

With all this going on I hardly gave my parents a thought but the first term was soon over and with my bag of books and dirty laundry, I caught the train home for 3 weeks respite. Without internet or mobile phones my main contact with home was through a weekly letter, one of those with stamps on. The communal telephone by the front door of the Hall was never answered and anyway long-distance trunk calls cost a fortune.

Christmas and New Year at home was a lovely break but I missed my friends in Liverpool and the intellectual rigor of our conversations as well as all the in-jokes. Somehow my old friends at home seemed rather staid. Young farmers or men training as accountants and stockbrokers, girls on secretarial or modelling courses all seemed rather dull; what a snob I must have seemed. Anyway, I was glad to get back to Liverpool for the new term.

---

[5] Bill Brockbank and Frank Dye sailed a Wayfarer dinghy from Scotland to Norway in 1964 surviving appalling weather and several capsizes.
[6] Gybe. When a yacht changes direction so its stern crosses the wind and the sail and boom crash violently across from one side to the other.
[7] Planing. Small boats when going quickly rise up on their bow waves allowing them to accelerate significantly.

In addition to the ongoing dissection and the lectures in anatomy there were other subjects including embryology, histology and organic chemistry. This last subject was taught in the Veterinary Building with twenty or so vet students sharing our very boring lectures. In protest at the quality of the lecturer the vets drove a herd of sheep into the class one afternoon. Despite the loud baas and the strong ovine smell, the lecturer droned on as if nothing unusual was happening.

The end of year exams eventually arrived and most of us passed, but not me. I had somehow failed anatomy by a small percentage. While my friends went off for a carefree summer I had to stick at my books for another few weeks, there was a crash refresher course and the individual attention I received from a senior lecturer opened my eyes on how to study, not as a schoolboy but as an adult. From then on, I gave up the constant re-reading of my textbooks and learned how to use the library and to read round a subject and see it in its full context. My failure was certainly a wake-up-call but much more importantly I came to think more deeply about what I was being taught and to try to fully understand the fundamentals first, before cramming in the detail.

CHAPTER 4. THIRD YEAR.

Third year; the year numbers are a bit confusing since the medical course is nominally for six years but hardly anyone does the first year. Most students go straight into so-called second year. First year offers the chance for non-scientists to enter medical school, it gives them four years of school-level science crammed into three terms; it's only for the truly determined!

So, into 3rd year; anatomy can now be put behind us as a bad dream. We will study function now that we have been taught the normal structure of everything. The great Prof. Gregory, of Gastrin fame, was our chief teacher in physiology. We experimented on pithed frogs; (at least we didn't have to do the pithing ourselves). We observed how the poor frogs' hearts reacted differently with changes in the concentrations of salts in the solutions in which they were bathed. We also experimented on one another; nothing very dramatic, we were not little Frankensteins, though we got dizzy blowing into tubes or half-blind for the rest of the day from staring into bright lights. Then we laboriously boiled up test-tubes of Benedict's Solution with urine for half an hour to test for diabetes and then were shown how to use a paper dipstick to get the same result in a few seconds.

Biochemistry and neurophysiology were hard work, though the latter proved to be more straightforward once I was able to interpret our lecturer's misuse of English. He was a brilliant man who had escaped Hitler's clutches nearly 25 years previously but he had not lost his broad German accent and it required more concentration to understand him than I usually could muster before breakfast.

I was still living at Rathbone and now had more free time without the tyranny of the anatomy course and I was determined to make good use of it. Having made friends with the new captain of sailing I was selected for the University first team after defeating most of the other competitors in the selection trials. Our team of three helmsmen and three crews proved a strong one and we were undefeated in our matches until the final events at the Universities Championship on the Welsh Harp, near London, when we were ejected in the second round by the eventual winners.

My regular crew was a music student and budding concert pianist Chris Welles, he was very dashing, his flowing hair worn long like the young Liszt. He was a perfect sailing companion with a good mind for tactics. Constantly on the lookout for opportunities like wind-shifts or the moves of our competitors, he made sure I steered the right course, while I could concentrate on making the boat go fast. Chris was being treated for epilepsy; he'd had a few fits since puberty and although he was supposed lead a calm life and drink no alcohol, he in fact lived life to the full. He had also been advised not to do any adventurous sports, especially those on water. Here we were then, aboard a 12-foot Firefly on a breezy Sunday morning in November on the choppy brown water of the Tyne Estuary. Liverpool had won the first race of the day beating the Newcastle University team. Now in race 2 Liverpool had won the start and Chris and I were lying second covering the best of the Newcastle boats, we were just to leeward and ahead of her. As expected, she tacked away and we went with her intending to slow her down and allow the leading Liverpool boat to get away into a safe first place. As we approached the first mark Chris said he felt a bit odd and then he stiffened and almost went overboard, but I grabbed his collar and forced him down into the bottom of the boat where he lay convulsing in every limb and making strange noises. I waved to the rescue boat but then decided it would be safer to just get him back to land, and so I hauled in the sheets and trimmed the sails heading for the beach on a broad reach. We pelted back, planing on every wave, Chris lay in the bilge-water now noisily snoring but looking a better colour. I could do nothing for him as all my efforts had to be directed towards keeping the boat upright; if we had capsized Chris would have little hope, despite his wearing a life-jacket. The hosts on shore wanted to call an ambulance but Chris, now recovering, forbade it. This experience rattled me and I considered asking for another crew, however Chris was very contrite and admitted he had been irregular in taking his Epanutin capsules and promised to act more sensibly. We finished the season together and then he left Liverpool for a conservatory somewhere on the Continent. I kept in touch through his girlfriend, a fellow medical student, and I learned that he had died suddenly a few years later.

Back on land I was shanghaied one evening by the debating society as a convenient medic to second a motion about some aspect of medical ethics. It seemed that I made a good impression and so was invited to speak a few more times. I started taking an interest in student politics and I was asked to stand in for one of the members of the entertainment committee at the Union who had resigned for some reason. This was very exciting as I met several pop stars of the time including The Who; they even smashed up their guitars at a concert I had to supervise. This destructive behaviour on behalf of heart-throb musicians is not original, apparently the young Franz Liszt had been notorious for breaking grand-pianos during fits of passionately performed arpeggios.

Towards the end of the year some of my fellow residents at Rathbone asked me to stand for the Presidency of the Hall's student-committee to try and prevent the election of an unpopular candidate, one Richard Calvert, a man whom I quite liked though he was perhaps a bit rough round the edges. I won the election with a big majority and as a result enjoyed my third student year greatly. I was automatically invited to many official functions, dances and dinners as the elected representative of this large body of students. One of my regular duties was to say grace before the Hall's weekly formal dinners when we all had to wear our academic gowns. My gown as president was sumptuously adorned with green silk while the other undergraduates wore their plain black gowns. The professors on high table sported brightly coloured gowns often with fur trimmings. I was proud of my atheism at the time and having to say grace rankled a bit so I usually rattled off some sort of dog-Latin with words like 'dominus' or 'Christay' before we all sat down to feast. On one occasion the guest of honour was a don from the Oxford University medical school. After I had intoned 'Levator palebrae superioris alequae nasii, amen' he leant over and observed that I had just named a tiny muscle in the face. My fast-fading anatomical knowledge did have its uses sometimes.

Being a student politician was not without its problems; one escapade landed me in the pages of the Daily Mirror and threatened by the University disciplinary committee. The student body of Rathbone Hall had a constitution but it was out of date and the auditors insisted that it had to be changed to comply with some change in company law; this needed a

50% quorum at a general meeting of residents. I made several attempts to gather the necessary 135 students to vote on this without success; nobody was interested.

Rathbone Hall: Student president chatting up Mrs Coult, the warden's wife.

Then I hit on a scheme; we would show some blue movies, but beforehand they had to vote. I got my quorum and a unanimous vote in favour of the amendment, after which some ridiculous flickering black and white films were shown with titles like Nudes in the Snow, Suzy's Bath Night and a third with a bevy of undressed ladies being chased by a man in a gorilla suit. One of our number, a budding journalist, wrote this up for the Daily Mirror and made it sound much more lurid than it was and I was notorious for a week or so.

CHAPTER 5. LET LOOSE ON PATIENTS AT LAST.

Years 4, 5 and 6 are the clinical part of the course. There was still plenty to learn in lectures and seminars and also in the lab, as we peered down microscopes at mystifying coloured slides of different germs or biopsies. We came to distinguish between leukaemias of different types, and could pick out a tubercle bacillus or a cancer cell. But most of our time was spent 'walking the wards' in a clean white coat, our pockets weighed down with notebooks, patellar hammer and retinoscope and wearing a Littman stethoscope casually around the neck, just like real doctors. We were actually interacting with patients. Of course, we were not allowed do anything beyond taking a history or making a basic examination but that was quite enough to stretch our abilities at the time.

Eight students became part of a medical or surgical firm which, in descending order of status, was comprised of the consultant, registrars (senior and junior); SHO and houseman but whatever anyone said, the most important person there was the ward-sister. She controlled everything and without her cooperation nothing was done properly. The consultant could stand up to her if he dared but we juniors had to work with sister. If she approved of you, life was simplified, but if she did not, then God help you. Each student was allotted several patients and we were supposed to know all about them from their arrival in the ward until they left; some going home, others to the mortuary.

We assembled outside sister's office, where through the glass we watched and waited as the doctors chatted and smoked over their tea. When they had eventually summoned sufficient energy, we all, in order of importance, followed the consultant to the first bed where a neatly spruced patient sat to attention in starched and unwrinkled bedlinen.

The consultant, Dr Watson on that ward in the Northern Hospital, shook the man's hand and addressed him by name; he had seen him before in the out-patient department. 'Who knows about Mr Smith?' he asked. A nervous student put himself forward. Usually, he knew everything in the case-notes and would start to trot out the facts he remembered, but sometimes he had to give the answer, 'Um, Mr Smith came in last night and I've not seen him yet'. Either way he would be roasted for some detail he had missed. In actual fact, Don Watson was a decent man and kind to both students and to his patients, though many consultants were kind to neither.

We spent our hours in the wards, with the registrars taking turns teaching us 'clinical method' which had to become habitual behaviour when we faced our patients. We would be tested on it, and how well we used it would be assessed in our final exams and again and again in specialist exams for Memberships and Fellowships of the Royal Colleges. Clinical method consists of listening carefully to what the patients say when we ask a series of open, and then later on, leading questions. There is an order to this: first we have to find out what the patient is complaining about, why he is in hospital and why someone called the ambulance for him. The doctor may have an idea about this but often the patient surprises you with a completely different scenario. You then need to discover everything about his illness; symptoms as they occurred over time. That sorted, you asked about his age, his work, his family, and what anyone might have died from. Does he smoke or drink or have any other interesting habits? Does he keep pigeons?

When they related their histories, I realised what hard lives many of our patients had lived. Often their relatives had succumbed only a few years earlier to conditions that we now regarded as relatively benign. The advance in medicine and surgery in only a generation was starkly clear with the introduction of penicillin, anaesthetics, biochemical testing and radiology, better training of doctors, as well as improved housing and nutrition. A woman might tell you she had had seven children and buried three; this would be recounted in a matter-of-fact way, without emotion; that's how life was. Children died of scarlet fever or meningitis, 'blue-babies' with congenital heart disease survived only a short time, tuberculosis killed teenagers, and young women died in childbirth or from puerperal sepsis a week later. A grandmother told me of her five children: one had gone 'funny' after having a high fever and went into a home, another had died from a burst appendix, her husband was killed in the Blitz, but she now had several healthy grandchildren of whom she was very proud.

We needed to find out a patient's previous medical history stretching back to their childhood, then move onto the present problem. No leading questions are allowed initially as, wanting to please, some patients will adopt some of the more interesting symptoms that a doctor might suggest to them. This is then followed by a round-up of any bodily system that hasn't yet been mentioned, being sure to leave the reproductive system to last. After all this you should have formed an idea of the diagnosis and you can go on to the physical part of the job.

The patient must lie comfortably on his back in a good light, preferably from over his left shoulder with the doctor on his right side. As each part of the body is exposed you will use your eyes first and your fingers later, not forgetting to listen and even use your sense of smell. Through all this you have to be careful not to upset your patient or hurt him. You need to protect his dignity, to explain what you are doing, and always look confident. Warm hands are always a good idea.

You are looking for any departures from the normal, which is why you have spent so much time in the last two years learning what that might be. Is the skin a bit pale, a bit yellow perhaps? The heart sounds, are they normal? That lump you can feel in his belly, is that supposed to be there? It takes many patients before you can confidently say whether this is normal or that is not. When you have some idea of the diagnosis, you might write that down in the notes together with a further list of possible diagnoses you need to exclude. With this all fixed in your mind you can decide on what lab or X-ray tests might be helpful.

This is clinical method and it does take time. How then does a doctor manage when he has to get through several patients every hour? Usually the diagnosis is obvious, or quickly becomes so. Much of the method can be dispensed with in a busy clinic though we must keep in mind that the diagnosis is only a 'working' diagnosis until we have confirmed it. This might take special tests or an exploratory operation, and sometimes it's a post-mortem diagnosis; occasionally one never knows as the patient annoyingly recovers before you can pin it down. However, clinical method is the key.

At one hospital where I worked, someone had stuck up a sign in the doctors' changing room. 'If you don't know what's wrong...go back and examine the patient again'.

For the next three years we honed the clinical method and learned a whole lot more besides.

I cannot recall my very first patient, though since our boss was a thyroid specialist, he or she was probably suffering from an overactive thyroid gland. The Northern Hospital stood near the city centre among terraced houses, dingy pubs and corner shops, derelict factory buildings, and bomb sites; all pretty normal housing for Liverpool at the time. It would have been called a slum anywhere else. The local denizens were poor and seemed to have been badly served by the local medical services although the Welfare State and the National Health Service had been providing free medical care for 20 years. People arrived in our clinics in the late stages of treatable diseases. Many were absolutely virgin cases untouched by healing hands. We wondered at their stoicism and wondered whether the local doctors were completely incompetent and had not noticed the significance of their complaints. Whatever, they provided the necessary examples that we students needed to learn our profession and we were duly grateful.

Arrogant? Indeed, we were; and oh, so proud to be part of the hospital system such that we reviled the non-specialist family doctors who always seemed to be getting it wrong while we, or at least the hospital, would cure where they had failed.

In an attempt to disabuse us of these notions, the medical course included the then innovative idea of student attachments to inner-city General Practices. Counterintuitively, these visits to their surgeries only confirmed our poor view of the unscientific approach taken by the lowly arm of the profession which in our biased view was staffed only by those who had failed to do better and gain a hospital appointment.

Bill had recently married Sue, a teacher, and I had been the best man. They had taken a first floor flat just across Sefton Park on Marmion Road in a big old house which had fallen on hard times. I often went over to see them in the expectation of being fed. Their upstairs neighbour was a rather alarming Irishman of great size. One evening I was with them watching TV in front of their gas fire, when the Irishman burst into the flat and attempted to attack the TV set, pulling the cables out of the wall, all the time crying, 'The rays! The rays! They've got into my brain'. We calmed him down and sent him back upstairs, pleased, but a bit scared to have met a genuine paranoid schizophrenic in its natural habitat. In future we watched programmes with the sound turned very low.

My brother Andrew was in his second university year studying metallurgy and engineering. He too lived in Rathbone. It was great to have him around again. He was an excellent dinghy crew and we had sailed together as a team since we were youngsters. With Chris Welles' departure to Vienna, he started sailing with me on the Liverpool team which continued its conquering career for a further year. At the end of fourth year I had to give up sailing for a while, as now clinical responsibilities were giving me little time off, nevertheless I was awarded full sports-colours for my two years efforts on the water.

As time went by, we students were allowed more and more clinical responsibility on the ward and soon we were taking blood specimens, putting up drips and assisting in operations. I haunted the emergency ward as, with luck, the casualty officer might let me do some bandaging or suturing or even set a minor fracture. One night we were in Casualty when sister told me she had just the patient for me, by which I understood it was yet another weirdo. This one stood there looking surly in his long dirty raincoat, his skinny legs and bedroom slippers just showing below its enveloping folds. He regarded the available staff in petulant despair: three women; the doctor, sister and staff-nurse plus me, the only male and it was to me he pointed. 'Im. I want only 'im and none of youse women' he announced. So, I led him off to the examination room and pulled the curtains. Opening his mac, he was revealed in his glorious nakedness, his tumid penis stuck in the ring-shaped handle of a metre long metal fire-poker. What now? I had few ideas and could only think of calling the fire-brigade to bring a metal-cutter; however the casualty officer knew exactly what to do and in 30 minutes she had released him from his burden. Her method involved wrapping thin twine tightly round and round the penis starting at the tip. The twine was then pushed through the obstructing ring and then gradually pulled to unwind it as the ring was peeled down the now compressed organ. It does work, you might want to try it yourself!

Casualty was sometimes a dangerous place with violent and drunken patients and their friends causing occasional mayhem. Fortunately, there were usually a few policemen hanging round drinking our tea and eating our biscuits in the dark hours. It's no surprise that casualty sisters often married policemen in those days, and maybe they still do.

One late night I was quietly stitching up a head wound on a comatose drunk when he woke up, saw me, swore, grabbed the scalpel from the sterile tray and had a go at slashing me. The reactions of the student nurse assisting me with the procedure were quicker than mine. She grabbed the instruments and dragged me out of the room, locking the door from the outside. The noise of the patient shouting and banging around brought one of our uniformed guests out from the tea-room. 'I'll get me dog; she'll sort him out' and indeed she did. In hardly a minute there was silence behind the door and the patient was then quiet and ready to let me finish the stitches with a few extra for a dog-bite on his wrist. I expect that nowadays this incident would result in all sorts of investigations and paperwork, at the time the patient would have felt lucky to only get bitten.

Later on, one summer afternoon, a policeman rushed into the casualty ward with a huge Doberman in his arms, unconscious. He and the dog had been playing fetch with a squash-ball in the carpark, when the dog had caught the ball from a full toss and it had lodged deep in its throat. With the dog on its back on a trolley, I reached in to grab the ball with some long Cheatel forceps. As he took his first breath, the dog's glazed eyes cleared and with a roar he leapt at me. I was quicker that time and ducked under the trolley as the great dog bounded outside, chased by a happy and very relieved constable.

We spent our days clerking patients, sometimes helping with bed-making and once, only once, I carried an enamel bed-pan to empty in the sluice. The registrars, 5 or 6 years ahead of us in their careers, were always ready to give teaching at the bedside. One afternoon our firm was standing round a patient's bed. He was due for surgery on the morrow. He had a swollen scrotum; James, our registrar wanted to demonstrate how we should examine it correctly. The patient lay there, his enormous scrotum displayed, as we crowded round behind the privacy screens. James had shown us various methods of proving this was a fluid collection within a cyst rather than a solid tumour, and he then went on to demonstrate the phenomenon of transillumination. If a penlight is shone through a solid mass then no light penetrates; on the other hand a fluid cyst glows brightly. As it is difficult to see this in bright daylight, he showed us how to use a cardboard mailing tube pressed against the cyst; if you peer down the dark tube the glowing fluid can be easily seen. We all performed this feat, which nowadays in the era of ultrasound seems archaic, though it is quick and costs nothing to do. One of us however wasn't paying attention; Mike, a good-looking young man with wavy dark hair and the glowering eyes of a Heathcliff, was idly watching the staff nurse working at the bed opposite. James called him back to this world and handed him the torch and the tube Completely mystified, Mike placed the tube on the swelling and applied his ear to the other end. After a few moments of concentrated listening, he announced 'testicular sounds are normal' and handed back the tube. Result; as Flann O'Brien would have said: 'consternation, laughter'.

Time was passing, we had completed three years of the course. Most of my non-medical friends were now leaving university for the great outside world to begin earning their daily bread, marrying their girlfriends and taking on mortgages. We medics still had another 2 years to go. For me it was time to leave the sheltered life of Rathbone Hall and start looking after myself and move into a flat. Sefton Park was surrounded by fine old houses along its 'drives'; some were still family homes but mostly the well-to-do had moved further out from the scruffy and petty-crime ridden city into the smart suburbs on the Wirral with their clean air and views over the Dee Estuary towards Wales.

Most of these old mansions round the park where rich merchants once dwelt surrounded by servants and large families were slowly deteriorating; a few derelict and abandoned, some split into flats or single room bed-sitters. The row that stood along Mossley Hill Drive had been acquired by the University to house staff and students, there was a plan to eventually redevelop the area into a new campus.[8] The biggest house, Number 1 housed university staff, number 3 was derelict, number 4 was for architectural students and number 6 medics. I moved into Number 4, the only medical student in the house. I had quite enough of my fellows all day and didn't need to live among them too,

I could see that Number 4 had once been a fine home with big rooms at the front and a billiard room on one side. At the back were the kitchen and staff room near a wonderful butler's pantry with a teak-wood sink and glass fronted cabinets. Up the narrow back stairs on the top floors was a warren of smaller rooms under the slope of the roof, all with large dormer windows. The building felt cold and damp, a smell of rotting timber pervaded the air, a generation of junk-mail lay in the cavernous hallway; I immediately loved it and set about decorating my room on the first floor.

Number 4, Mossley Hill Drive.

---

[8] Revisiting the area in 2019 after an interval of over 50 years I saw that all these houses had been tidied up and were occupied as two-or three-bedroom flats. A resident told me that as they were grade 2 listed, the houses couldn't be demolished for the originally planned campus development.

Floor boards were stolen from the derelict House Number 3 next door to make bookshelves and a desk, the Smithdown Road auction room provided two arm chairs and a mahogany wardrobe. My mother promised to buy me a new bed, so long as it was a single one; what did she suspect? An old carpet from home stopped some of the drafts from gaps in the floor. There was already a modern gas fire in the marble chimney-piece and a junk shop provided a little gas ring for making tea by the hearth. I had my own wash-basin in a corner and two huge sash windows looked over to the park, where I could see the top of the palm-house roof above a line of trees. Perfect; I went on living there for several years, even after I qualified, although the hospitals provided their own accommodation for the housemen.

Fellow residents at Number 4. Crowther top left and Charles just below him.

Across the landing lived Charles, he had left Oxford sooner than he might have hoped after spending too much time at the rowing club, and was now studying physiology under the great Prof. Gregory; he had my sympathy. We immediately became friends. Another denizen of Number 4 was a post-grad biochemist, one Geoff Crowther, a very bluff Yorkshireman. He lived with his cat on the top floor and talked incessantly of his wish to travel to the exotic places of the world. One day he loaded all his possessions into his ancient Morris 8, and with the cat's basket in the front seat, he set off on the drive to India. We all waved him farewell.

That afternoon he returned to us after breaking down on the Runcorn Bridge. He did eventually make it to India and to Africa, and as history relates becoming one of the founders of the Lonely Planet Guides.[9]

Number 4 was infamous for its weekend parties, all held in the ex-billiard room. On Saturday nights the architects, tired after a week emulating Christopher Wren or Frank Lloyd Wright, liked to relax with loud music, wild dancing and plenty of lager. Dazed and scantily dressed girls wandered round the house looking for the toilet or perhaps a spare man. Around 4 am silence fell. All week our coin-box phone rang. Occasionally someone answered; 'Is there a party on Saturday?' the caller would inevitably ask. Our number was easy to remember; Allerton 3825. Just dial ALL F**K, it fitted.

The bathroom next to my room had been disused for a while, in fact ever since the drains got blocked; I called in the University's handyman to sort it out. I arrived home early one afternoon and found the plumber holding a length of eight-inch-pipe completely occluded by a horrible rubbery mass. 'If you lot spent less time with girls and more time at your studies......'. Well, I did say we were growing up.

Although we often had a take-out from the chippy near the bus-stops on Smithdown Road I preferred to cook my own fresh food in Number 4's enormous kitchen. I had by now progressed slightly beyond Katherine Whitehorn's excellent book *'Cooking in a Bed Sitter'* and was trying my hand at more exotic dishes. Food in the hospital canteen was filling but stodgy and I couldn't afford the prices in good restaurants so to satisfy my tastes I had to cook for myself. Girlfriends took their lives in their hands when they accepted my invitations to dinner *a deux chez-moi*.

---

[9] Geoff Crowther who died in 2021 was almost a deity during the 1970s and 80s to backpackers in the wild parts of Asia and Africa, his writings guiding them, often inaccurately, to the remotest outposts beyond civilisation. An interesting man whom I knew, but never well, for almost a decade.

CHAPTER 6. STUDENTS INTO DOCTORS.

Fifth year started with 3 months at Broadgreen Hospital out on Queens Drive, the main inner ring-road round the city. Travelling between the university campus, Mossley Hill and far-flung hospitals was becoming time consuming and involved changing buses. Ignoring my father's advice, I cashed in my Premium Bonds and bought a second- hand Austin Mini. It came to live in Number 4's driveway next to Charles' Triumph sports-car of which I was very jealous. I was not green for long as Charles soon wrote it off in Snowdonia, attempting the notorious Ugly House corner at too lively a speed.

The lectures continued. Every Saturday morning at about 8am we were expected at the 'Meat-Class'. Each week a few of us were chosen to perform. Lumps of human tissue that had been removed from living patients in the operating theatres or from the dead at post-mortem were exhibited. A student then had to describe what he saw and make a stab at diagnosis. The clinician responsible for the case usually turned up to explain the history and diagnosis and this was followed by lively discussion. Though a bit scary it did give us insight into disease management and introduced us to some of our future seniors.

We were also subjected to the weekly 'Circus' at the LRI. One or two patients with interesting problems were brought down to a lecture theatre in the basement of the hospital among the serpentine pipes snaking along the low ceilings underground. A student was allotted the task of taking a history during the afternoon of the Circus and he would present the case to an audience of professors, visiting luminaries, hospital staff and fellow students. The cases were always interesting and presented problems of diagnosis or management, the occasional cases were just oddities. When it was my turn, the patient was such an oddity, he had been admitted for some minor surgical procedure but was found, quite incidentally, to have dextrocardia; he had been born with his heart on the right side of his chest. To prolong the mystery for the audience I posted the chest X-rays on the stage's lightbox back to front and then reversed them for the denouement. I got off lightly as there was little to tell, others struggled with rare and complicated cases. We were learning that you never knew what would turn up next.

After the New Year in 1967 our firm went to Mill Road, until a few years earlier the largest Maternity Hospital in the country. We were to spend 2 months doing gynaecology, then 2 months in the delivery wards under the gentle hand of *Mister* Basu[10] the senior registrar to learn basic obstetrics. Basu had a terrible reputation among the students as a slave driver. During the two months of our maternity course we were not supposed to leave the confinement of the hospital, not even to go to lectures back at the medical school.

Sleep was in very short supply since, as is well known, babies prefer to arrive during the hours of darkness. At the same time we were bashing the books for the final exam in Pathology which was almost as difficult as Anatomy, with its myriad small details. This exam coincided with our last days at Mill Road and it was accepted by the University that this coincidence was a hardship for us. Mr Basu however didn't consider this an excuse for any leniency on his part and he drove us just as hard. We were by now old hands and knew that to handle the likes of Mr Basu one had to acquire a reputation as a hard worker and an early-riser starting from day one; once you achieved that you could pretty well sleep all day undisturbed. This worked with Basu and after a week or two of our running *him* ragged he relaxed.

There is not much dignified about gynaecology, we try to make it so, but for the patients it's inevitably an embarrassment submitting to the intimate examinations necessary. As students we were well aware that our presence added to the women's unease. We watched the specialist and then we tried to emulate his skilled touch of hand and then, using a speculum, to examine the cervix and the rest of the vagina, to take swabs and smears. I was very grateful to all these patients who perhaps reasoned that their helping in our training would eventually benefit their fellow women sometime in the future. Eventually as the course went on and after a few weeks I stopped thinking about it this way and, like most of the

---

[10] In Britain surgeons are called 'Mister'. After all those years working to become Dr So-and-so they then toil for another five years to be Mister or Ms again.

patients, accepted it for what it was; a necessary part of life, just get on with it. It was a skill we had to obtain. For the women it was a part of what life threw at them, living in a poor area of the city. Poor diet, poor housing, little education, too many children; it was their life just as it had been their mothers' and grand-mothers'.

We lived in the students' residence on a floor above the labour ward; I shared a room with Bill as we students settled in to the worn-out space with tattered carpets, collapsing arm chairs, squeaky beds and dusty windows. On our first evening I watched the delivery of a baby for the first time. I was amazed and shocked, how did people do this? The pain, the blood, the sheer physical effort. I wanted to stop it, to let the mother rest, to have no more pain, to stop bleeding. But then it did all stop, and that with a marvellous beginning; the start of a new life. The baby gradually emerged headfirst and was lifted up into the world; the mother smiled through her fatigue, the battle-hardened midwife congratulated her warmly, the baby yelled. Another life, everything else forgotten, mother was cleaned up and she lay there cuddling the child at her breast; smiling, proud and exhausted. I too felt the emotion, I had only been an observer, but I knew that the next two months would be exciting and perhaps life-changing for me.

Having watched just one delivery we were now to deliver 20 babies ourselves; we had to compete with the student midwives for babies as they had a quota too. However, the good women of the area kept up the supply. In Roman Catholic Liverpool the priests held sway, Rome disapproved of family planning and Mill Road churned out as many souls to be saved and redeemed as the Pope could wish. Night and day, whatever we were doing, the head midwife would call for the next student to come to labour ward. We were closely supervised for the first few labours but soon we were working on our own, encouraging the mother, trying to control her pushing and her breathing as we'd been taught. We observed the mother's vital signs, listened for the baby's heart rate with an old-fashioned conical stethoscope, estimated any blood loss and filled out the chart clipped to the end of the bed. Progress was checked by the degree of stretching of the cervix and we became good at estimating this by touch; as dilatation reached its maximum we wheeled the bed into the delivery room. In a first-time mother, a so-called 'primip' there could be a long delay now, in a 'multip', especially a 'grand-multip', the baby might arrive with a rush.

One of our chores was the monitoring of the women whose labour was being induced. For a number of reasons labour might need to be artificially started; perhaps because of high blood pressure or maybe the placenta was failing and the baby's growth had slowed down. The older midwives reminisced about the days of the Liverpool trams, scrapped some years previously. Anyone arriving by tram would be in labour in those days after a jolting journey up the hill to Mill Road, the modern apple-green Leyland diesel buses just had too smooth a ride.

First the amniotic membrane is slit, releasing the 'forewaters'. Syntocinon is then infused by an IV drip in increasing doses until contractions start. For hour after hour a student sits at the bedside checking progress. The rule was that 'the sun should not set twice on ruptured membranes'; if induction failed then delivery by caesarean section was inevitable. Mrs Andrews was a woman in her 20s having her third child, she was very obese and her rolls of tummy fat made it hard to feel the uterus at all, even to determine whether she was having contractions. I had sat with her all night, for about 18 hours and we were both exhausted but nothing was happening; no contractions, no baby.

That was the morning of the weekly 'Grand Round'. Professor TNA Jeffcoate, the author of the great textbook, arrived with his entourage. They all trooped down the ward in order of precedence; visiting professors and lecturers from famous universities first, then junior consultants, obstetricians in training, Mr Basu, the houseman and medical students somewhere at the back, their echelons all arranged to the right of the patient. On the other side of the bed were the nursing staff from Matron down to pupil midwives, all in their elaborate bonnets and starched pinafores, badges gleaming. Everyone on parade. All here to honour the author of the great textbook, Professor TNA Jeffcoate with his brilliant white coat and his elegant silver hair.

They bore down on Mrs Andrews and me. Her case was quickly discussed, 'failed induction' pronounced, caesarean section ordered this morning. Everyone was ready to move on, except for me. In my sleep-deprived state I chose to disagree. 'No, wait a minute'. Many pairs of eyes swivelled towards me in surprise. Upstart! Doesn't he realise the great Professor has spoken?

'Well,' said I, now feeling a bit abashed at my presumption, 'operating on Mrs Andrews won't be easy because of her obesity, her uterus is deep down in inches of fat and her post-op recovery is going to be complicated by her inevitable immobility; I feel it should be avoided if at all possible. Anyway, has she really had a proper trial of induction? She weighs more than 250 lbs and we have been using doses of Syntocinon suited to ladies of half that size'. Looking at Mr Basu, the Prof. asks whether Syntocinon is fat-soluble. Nobody knows. The dose is increased and Mrs Andrews delivers a healthy boy that afternoon without a Caesarean. Am I a hero? A leper more like!

Thus, the two months went by in a sleepless haze; babies yelling, mothers pushing, sudden bleeding, emergency sections, lectures on breast feeding and maternal nutrition. Babies were snatched from the jaws of death, most cried lustily, very occasionally one died or was born with a deformity. For us it was becoming business as usual; our patients came in, had their babies and left. Some would remember 'their student', most would not. I remember them all with gratitude and with respect for how they laboured and somehow kept their Liverpool sense of humour. Said one woman after having her fourth child as I put sutures in her torn vagina, 'stitch it all the ways up, luv. That'll keep the bugger out!'

The obstetrics course was probably the critical point in our training; I had started in the maternity wards as a student and in just two months began to feel like a doctor. I had taken responsibility for mothers and babies; my efforts had contributed to their care and their safety. After this experience there could be no turning back, my life became more serious and my approach to it had to be more mature. We were no longer just learning; this was not play; we stood at the border of a new world which we were almost ready to seize.

The time had come to move on and I emerged from Mill Road, where we had been cooped up for eight weeks, almost blinking in the bright light of outdoors. The Pathology final exams would fill the next week and most of us had already logged our statutory twenty deliveries. Bill piled into the Mini with me and with books and suitcases we drove back to Sefton Park. I dropped in to the Marmion Road flat to see Sue who was large with child and due anytime soon, booked into the other Maternity Hospital on Oxford Street near the Medical School.

The written papers were soon done and the last act was to be the viva-voce exam. Visiting professors and dons grilled us over microscope slides, preserved bottled specimens and the raw meat removed at recent post-mortems or surgical operations. Mine was all over before lunchtime, I met Bill for a sandwich in the Union; he was due in the exam room about tea-time and enquired anxiously about what I had been asked. Of course, it was too late for last minute swotting now, but nevertheless I sat with him as he flicked restlessly through Boyd's Textbook. I stayed with him for I knew that Sue had just started in labour and had been admitted to the maternity ward. It would be several hours before she would deliver but we all agreed with Sue that Bill should be kept in the dark until he had finished the exam. Weatherly, one of our more unworldly fellows, was sent to accompany Sue and had helpfully packed her a suitcase. It only contained a pair of jeans, one of Bill's jerseys, several paperbacks and a skimpy bra.

Bill emerged from the exam room looking a bit goggle-eyed after his inquisition and we whipped him off to the Maternity Hospital. The midwives did not welcome the arrival of 20 enthusiastic medics and we were all sent home leaving Bill and Sue to get on with it. Baby Emma arrived after midnight, beating the midwives to it and Bill, by default, had the pleasure of delivering his own daughter.

Everyone passed Pathology, a few with distinction. The maternity course had been a tough few weeks and we were all changed by it, certainly I was. We had been let loose on patients with minimal supervision, we had been given a taste of the responsibility we would carry for the rest of our careers.

There was to be no rest, we were immediately back to attending daily lectures and started to learn some of the smaller specialties now we had done the three big ones: medicine, surgery and maternity. Now for everything else. Dermatology; the specialty where one is never disturbed at night. Anaesthetics, Public Health where we stared down drains, Psychiatry which proved very dull as we only saw a few neurotics and not a single maniac, Paediatrics and so on.

My first specialty was in the orthopaedic ward at the LRI. Orthopods tend to be regarded by other medics more as artisans than doctors. The chaps who got into medical school because of their prowess on the Rugby field rather than in the exam room often went into Orthopaedics, a calling that needed physical strength and stamina for manipulating arms and legs into position, setting fractures against the pull of powerful muscles or just standing for long periods holding a limb in one position while a colleague busily cuts and hammers away. Saws, chisels, hammers and power-drills were used in operations as well as the usual scalpels and forceps. It seemed great fun and when the junior houseman went off sick for a week, I volunteered to stand in for him; my first 'locum'.

Disappointingly I was not to work in theatre because the trainee surgeons on the MChOrth[11] course, paying guests as they were, assisted at the operations. However, I was allowed to reduce some fractures while I was there and became quite a dab hand at applying Plaster-of-Paris splints, even coating them with fresh egg-white to add a golden sheen to my masterpieces. I had to provide night cover for the ward every second night, a split responsibility with the houseman in the gynaecology unit, so I also had to deal with emergency admissions from Casualty to the women's ward.

My fracture patients and post-ops were mostly under control out of hours but the gynae ward was another matter. We were near the centre of the city and it was before the legalisation of termination of pregnancy. The Abortion Act came into law in 1967 and eventually put an end to the appalling business of illegal abortions carried out in unhygienic conditions by back-street abortionists. Who were they? Struck-off doctors or retired midwives perhaps at best, at worst incompetent and ignorant meddling women in filthy kitchens. The result of their activities was a steady stream of very sick women with blood or stinking pus pouring from their vaginas.

---

[11] Master of orthopaedic surgery, a necessary qualification for advancement in the specialty.

Often, they had septicaemia and high fever. Some died, many would no longer be able to have a baby. My job was to get them into the ward as quickly as possible, taking blood samples, setting up drips and ordering blood from the transfusion service. I was just doing what the duty registrar ordered, many patients had to go to theatre to complete the abortion or to stop bleeding and remove dead and infected tissue. For the first time I saw gas-gangrene, something I would never expect to encounter except perhaps on a battlefield. But sometimes for the very deprived, Liverpool must have been a battlefield, especially for a poor and single woman finding herself pregnant and with no support from her family.

As we moved between the hospitals and clinics in the city and especially when having teaching sessions in the peripheral hospitals, we got to meet a different type of senior doctor. Unlike some of the god-like consultants in the teaching hospitals the men and women whom we were encountering now appeared normal and very approachable, and they seemed delighted that we sought their knowledge. These were the people to whom we would soon be applying for our first jobs after we qualified.

BB was undoubtedly one of the gods but also proved to be very approachable. Dr E.T. Baker-Bates was coming to the close of a distinguished and perhaps controversial career. An impressive white-haired man in his 70s, always immaculate, he wore a morning suit in the wards all week and on weekends would appear in tweeds and plus-fours. He was said to be manic-depressive and heaven help us all when he was in a manic phase. When depressed he retired to his huge house on Rodney Street[12] to be isolated from the world by his dragon of a house-keeper. I first met him when he kidnapped me from an adjacent clinic to help him with his out-patients one blazing hot summer afternoon. All the windows were open but it was oppressively sultry in the consulting rooms. BB had chosen to remove his pin-striped trousers to keep cool. His long starched shirt-tails hung below his knees but his skinny calves in grey silk socks looked a bit alarming. Sister was beside herself and despite her efforts to confine him behind his desk he strode round the clinic and out into the waiting room too.

---

[12] Rodney Street, Liverpool's Harley Street.

'Doctor is in a state of undress' she informed every patient as they came in. Since most of his patients had known him for a long time they were none of them surprised or shocked; it was just BB. He and I became friends over several years until eventually I left England. He was an excellent teacher and a good physician. Despite his age he was always up to date with new developments. When making a point he would say, 'such and such, isn't it boys and girls? Isn't it?' He laid emphasis on close observation of the patients. Full of stories, many of them defamatory about his enemies of whom he seemed to have many, for he could be waspish. He was originally from a very modest background in St. Helens but had known Sir Thomas Beecham the conductor, also from that town. Despite his high regard for Beecham's musicality his stories about him were quite scandalous.

A cartoon of BB in his prime.

One of BBs aphorisms was that though many people visited a patient's bedside the doctor alone brought knowledge, 'isn't it? boys and girls, isn't it?'. I took this to heart and have always remembered it. All those others, the library lady, the ward cleaner, the man selling newspapers, visitors, the vicar; they all brought friendliness, concern, news from the outside world and so on, but we medics brought our knowledge. We should not squander our chance to help the patient with that knowledge by just having a friendly chat or sorting out his family problems and so on. We must be polite and respectful but our job was more than this, we were the ones who were there to get him better by applying our unique knowledge. Perhaps this sounds a bit a bit smug but it contains a kernel of the truth.

One evening Bill and I were invited to Rodney Street for a tutorial followed by a light meal; BB was fond of feeding his students; usually with his latest fad diet. As we arrived one of his middle-aged private patients, the lady-mayoress of a nearby town, smartly dressed with pearls and a nice hat, was carrying two full buckets of coal up the grand staircase. BB on the upper landing was encouraging her. We leapt forward to take the buckets. No! Leave her, says BB, she's got to realise how much extra weight she's carrying round. She must go on a diet. It was a good lesson; one has to be as firm with paying private patients as you are with those on the NHS.

St. Helens was a somewhat run-down and dirty Lancashire town in those days, dominated by the huge Pilkington's glass factory, yet still surrounded by the unspoilt farmland of South Lancashire. Near the town centre stood the Providence Hospital owned by a Catholic order of nursing nuns. BB was the consulting physician there and he liked to shanghai his students thence to do locums. I covered a few nights for his houseman and on one occasion, two whole weeks. I even got paid for this and was kept fed by the excellent hospital kitchens. I was the de-facto houseman despite still having a year to go as a student before Finals. I admitted patients, went to theatre to assist Mr Marsden the surgeon, wrote prescriptions, dealt with problems on the wards, set up drips and saw patients in Casualty. The nurses all recognised my lowly status and went to great lengths to help me and prevent me from making mistakes. I made friends with one of the younger nuns and would spend off-duty moments chatting with her over cups or sweet tea in the ward office while we waited for an X-ray or for the lab to send up test results. Finding our afternoons off-duty coincided we planned to go together to the nearby cinema to see the Sound of Music. As we arrived in the darkened auditorium and sat down there was great giggling in a row further back. Several of the senior nuns were also there to watch Julie Andrews and Christopher Plummer singing in the Alps, we moved back and helped them eat their box of Dairy Milk chocolates and afterwards had tea and buns in the cinema café. Not many of us get to go on a date with a nun!

Austin Marsden was a quietly spoken man and seemed rather introverted for that most extrovert of professions. He was an accurate and fast surgeon but very deaf and wore a hearing-aid that squealed loudly in his ear. Sometimes when a patient asked him a difficult question such as 'Is it cancer?', he would pretend to mishear and poke ineffectually at his ear-phone. In those days it was usual to keep the patient in the dark about a bad prognosis, the word 'cancer' was shocking and for some shameful. The patient was usually unable to summon the courage to ask the question a second time so this trick usually worked. One afternoon in theatre Mr Marsden had opened an abdomen and we were ladling out a slightly smelly gelatinous goo which had filled the peritoneal cavity, Pseudomyxoma Peritonei, a rare type of cancer. As he removed handfuls of this muck into a kidney dish he quietly muttered to the nurse, 'Sister. If it's jelly for tea; I don't think I will'.

Mr Marsden's great weakness was the St. Helens Rugby League Football Club. He had reserved seats for all their games and there was no hope of calling him on a Saturday afternoon. The club had moved to a new ground and the old site was sold for development. Mr Marsden had bought a large plot which had originally been near one of the goal-posts and there he had built his dream house. The open plan ground-floor was magnificent with its light oak flooring which was measled with little brass plates embossed with legends such as '5.3.1921 v Wigan: Bloggs' winning penalty kick' or '7.12.1948 v Huddersfield: Jones tackled Williams'. Each marking some historic event in a long-forgotten match. I wondered who kept them polished; probably Austin himself.

In the evenings I sat with the other doctors in our residence, a pleasant detached house in the grounds of the hospital. It was conveniently near the Casualty ward where I had to see patients when the surgical houseman was off duty. Sister Duffy was in charge of Casualty and she well knew my inexperience, incompetence? She called me on a Saturday evening to see a middle-aged man with a shoulder dislocation. He was a retired all-in-wrestler and had been thrown awkwardly while making a come-back in the ring at the town hall. I had seen dislocations before but never seen one being treated, so while X-rays were being arranged, I called the duty orthopaedic registrar, but he was unavailable and attending to the victims from a coach accident on the M6 at another hospital nearby. I would just have to cope! As, in near despair, I regarded the injured wrestler Sister Duffy called out, 'Dr Howard, Mrs Brown is looking for you'. This was code for 'come into the tea-room', where there was an enormous wooden lectern in the shape of an eagle with outstretched wings, just what you would expect in a nunnery supporting the Holy Writ. Open upon it lay a large and ancient tome, not the Holy Bible but the very next best thing if you are in a casualty department, and almost as out of date. It was the Textbook of Orthopaedics, 1898 edition, by the late great Sir Robert Jones[13]. Sir Robert recommended me to lay the patient on the floor and 'Place the socked foot in the supine patient's axilla and apply traction to the extended arm…. etc etc'. There was an accompanying engraving of a patient lying on the floor, the surgeon's foot in his armpit. Both men had fine Victorian facial hair. Excellent! I followed Sir Robert's advice to the letter and was rewarded by a satisfying sucking squelch as the dislocation went back into its joint. 'Not seen it done that way before, doctor' said the wrestler, putting on his coat and heading back to the Town Hall. I read on in the book to find, 'It is vital that the patient wears a sling for at least a week'. Too late to get him back now; I suppose he was all right.

---

[13] Sir Robert Jones, born in Llandudno in 1857, considered the father of British orthopaedics and perhaps the first to use X-rays in diagnosis of fractures

CHAPTER 7. FINAL YEAR.

The summer flew by; we only got a short break which I spent with my parents by the sea in North Wales. We sailed, walked, went fishing and I managed to read a few books that weren't about medicine.

My parents: Philip and Mary, on holiday in North Wales.

And so back to Liverpool, it was Final Year at last! We all learned how to anaesthetise patients and more importantly how to wake them up again. Inserting an endotracheal tube was a necessary skill for all doctors even if they never intend to become an anaesthetist. At the Dental Hospital, wielding special pliers I extracted two molars from a very brave young man. 'Thanks doc'; these people were always so grateful; the more it hurt the happier they seemed.

In the skin clinic we learned to distinguish different rashes. Since the treatment for all of them seemed to be exactly the same one wondered why. In radiology we peered at films in the dark and learned about exposure times and the dangers of giving our patients cancer, we gave smallpox vaccinations. At Alder Hey we endured two months of screaming kids being sick on us. My head rang from the constant noise of crying babies.

Then to the Sexually Transmitted Disease clinic, at the time more abruptly called the VD Clinic which was down in the bowels of the LRI. On our way to our first session, walking down the stairs Bill was holding the banister, 'I wouldn't like to be touching that rail' I say and Bill jumps away with a cry; we did plenty of hand-washing on that course. One of the VD clinics was down in the dock area where sailors from all over the world filed in. A ship had come in from Kobe with a dozen or so matelots with the usual symptoms. 'Show the pipe' commanded the doctor. Swabs and blood tests were taken and antibiotic injections given. Antibiotic resistance was becoming commonplace and penicillin was no longer effective in many cases; there was a particular problem with a strain of Gonorrhoea called Tokyo Rose arriving from Far Eastern ports. Although our patients were advised against alcohol and being friendly with the lasses of Liverpool until given the all-clear, such words probably fell on deaf ears and Tokyo Rose was spreading merrily through our city with every ship that came in. Contact tracing was important in containing outbreaks of VD and we students spent happy hours asking patients the names and addresses of their intimate friends. Not much information was gleaned under the circumstances.

One of the venereologists had a small fluffy dog who lay under his desk uncomplaining as the sailors went to and fro. He ignored the Lascars and the Africans, the few Europeans and Slavs, even the Chinese, but the Japanese he could not abide and flew into a yapping rage.

An Irish doctor, Mairead, asked me to cover her holiday and so I spent a fortnight as a locum houseman on X ward at Walton Hospital working for Doctors Skene and Alstead. They and Arun Baksi, the registrar, let me get on with it. Bert Alstead had emphysema and didn't like to climb the stairs to the ward so we would stay on the ground floor to discuss the patients, Bert chain-smoking as we did so. Sandy Skene was the superintendent of the hospital and didn't have much time for merely seeing patients. Arun was swotting for an exam and wasn't to be disturbed. I had never enjoyed myself so much in my life and got to know all the staff on the ward and I made up my mind to come back as houseman when I qualified. Walton was such a friendly hospital and the patients loved it too, it was part of the community in that rather deprived area of North Liverpool.

They told us that it was no longer compulsory to go to the David Lewis Foot Clinic which was housed in a building like a Victorian workhouse near the Anglican Cathedral, but nevertheless I went to see it for the cultural experience. It was due to be closed down in the near future, having been officially described as an outdated anachronism (i.e., It cost too much). All the tramps in England knew the Foot Clinic and faithfully dropped in as they trudged through Merseyside. Its job was to attend to the horrors that these men, and a few women, had at their lower extremities. Ancient worn-out boots had to be removed and layers of newspaper and old socks peeled away to reveal…, well what do you think? After an hour or so soaking in warm solutions their feet were ready for further attention, scraping, snipping, curetting. Some tramps had to stay some days while deep ulcers healed but most were given new footwear and sent on their way. The smell was indescribable but perhaps not as bad it could be in the post-mortem room, particularly when poor souls were fished out of the river.

The last months of the course dragged on with a week here and a week there as each minor specialty demanded our attention, so we visited this hospital then that; we attended lectures on arcane subjects like ethics and how to be a witness in court; we wasted a day hanging around a public health clinic waiting for someone to come in and be vaccinated and then Bill and I ended up vaccinating one another to get that ticked off. The ear, nose and throat lecturer spent two weeks of our time trying to teach us how to use a head mirror to examine inside ears, a hopeless task; it was just too difficult and in the end we preferred using a simple hand-held otoscope. At St. Pauls eye hospital I was allowed to excise a small cyst from inside someone's eyelid. This was all good fun but quite honestly, we were fed up with being students. We had walked the wards doing student locums and we wanted to be qualified doctors; five years as undergraduates was too long; we longed to get the exams over and done with.

I suppose we now considered ourselves beyond just looking and listening to our teachers, we wanted metaphorically to get our hands dirty. Apprentice bricklayers expected to actually lay bricks and mortar; but then a wonky wall is hardly in the same league as a mis-treated patient. We felt ready and fretted at being held back.

Then suddenly Finals were upon us. Under the lofty ceilings of a converted non-conformist chapel, we sat the written papers They were long but we were used to that; we had lived a life of exams starting with the 11-plus, Common Entrance, O levels and A levels, second MB[14] and last year, Pathology; so the written papers, while searching, we could handle. Clinicals were very different, we would be tested and questioned on our skills as doctors; history taking, examination, diagnosis, treatment. What's the dose of this drug? What does this X-ray show? Interpret these blood tests, this ECG. Test the urine for abnormalities. Look at this patient's eyes, hands, abdomen; describe this man's gait, this woman's tremor; what is this lump, this rash?

It took the examiners a week to decide and eventually little strips of paper appeared on the noticeboard outside the Dean's office. I didn't even bother to look at the list of distinctions, that was for a few others, but there I was on the longer list of names who had passed, and so were Bill and my other friends. MB ChB Liverpool 1968. Done!

To the red phone-kiosk outside in the street; a reverse-charge call[15] to my parents at home. Operator; 'Who shall I say is calling?' *'Doctor* Howard' I replied, the first time I was to use my brand-new title.

---

[14] Second MB exams at the end of 3rd year.
[15] If you are American, that's a collect-call.

BOOK 2. JUNIOR HOSPITAL DOCTOR.

CHAPTER 8. HOUSEMAN.

It was my first day at work as a real doctor, I was back at Walton Hospital and on X ward again; male medical. The summer of 1968 had probably been my last long holiday for a very long time to come. While in France rebellious students had been trying to burn down Paris I had spent two months in North Wales fishing and sailing, drinking and eating. I raced the family's GP14 dinghy in the Menai Regattas and won a few of the trophies. To my parents' disgust I swopped my faithful grey Mini for a two-seater Triumph painted in metallic gold. At the end of August, I returned to my old room in Number 4. Although the hospital provided accommodation of a sort, I knew I would need to get away when I could and be a normal person away from the frenetic activity of junior doctor life.

Two-seater Triumph painted in gold.

Every morning I picked up a clean white coat; the one I discarded would be creased and a little blood stained here and there, the right cuff[16] covered with ballpoint hieroglyphics; blood results, patients' names, clinic

---

[16] I am left-handed.

times, phone numbers. I was one of seven house-officers, all doing their first job as doctors. Joanna was my opposite number on U ward, the women's ward, on the floor below X. We alternated nights on duty.

When X and U wards were on 'take', local GPs, Casualty officers and Bed Bureau bleeped me incessantly. 'I have Mr So and so with a stroke/stomach pain/ heart attack/ you name it.' Where they were to be put, I had no idea; the hospital was always full. Other consultants, more on the ball than my two, kept their beds filled in order to block random admissions that they did not want whenever they were on take. They would not discharge patients until they had arranged the arrival of a new patient of their choice. Sometimes they blatantly lied about any vacant beds on their wards. But X-ward always had room for more old geezers with nowhere to go, whereas L ward was full of younger men with interesting conditions who will certainly get better and go home. So, X-ward was blocked up with old blokes with strokes, who, quite simply, nobody wanted. It was a rare victory when one of ours could be discharged back to his family, they mainly went off to the geriatric hospital, to care homes or to the mortuary. One family who managed to admit their grandpa to my ward actually said to my face that they would not take him back as they were sick of the old boy and they hoped to put a paying lodger in his room.

I was working for the two consultants whom I have mentioned earlier, both very nice men; Sandy Skene whose heart was really in administration and medical politics and Bert Alstead whose lungs, far gone with emphysema, hardly encouraged him to climb all those stairs up to X ward. When he did arrive, he usually sat in the office with the charge-nurse to recover his breath and smoke another cigarette rather than drag himself round the patients. I did not mind, I knew my job and I could usually persuade Arun Baksi, my registrar, to leave the library for a bit and check on what I had been doing.

X-ward was big and long: perspective made the rows of beds along each wall seem to converge when viewed from the door. About 50 men, mainly old and who had all had a hard working-life, sat or lay in their beds, a few in arm chairs wore dressing gowns. Others shuffled around chatting.

Many were puffing on a cigarette[17]. On each bedside stood a spittoon half filled with the ghastly green goop that almost every patient constantly expectorated, often with a fag-end floating atop. Those on oxygen especially should not have been smoking but many did. Some of the men had isoprenaline inhalers, not the new-fangled aerosol ones but old-fashioned glass ones with rubber bulbs that you had to squeeze. There was constant movement as nurses, assistants, junior doctors, ward orderlies and cleaners went about their business under the beady eye of the ward sister. Her immediate boss, the charge-nurse, would sit, bleary-eyed and bloated, in the office smoking and pretending to write up the day's records. He, like Bert Alstead, had emphysema; in fact, most of the men in the ward had emphysema. We called it Chronic Bronchitis then, nowadays it is called COPD, the result of a lifetime breathing industrial air, and cigarettes too of course, but then everyone has to have a ciggy, don't they? The tobacco industry, which was then busy trying to suppress the medical evidence against smoking, has much to answer for.

Jerry was in his late 70s but looked ten years younger, he had suffered a coronary thrombosis and would be with us for three weeks of rest in bed where he was visited by his new bride, they had only been married a few months earlier. He was a retired marine-engineer and had maritime stories to tell to anyone who would listen. He told me of when, just after the First World War, he was sent to Vladivostok to repair a ship's engine and travelled across Siberia by rail. From the train he saw processions of chained suffering prisoners walking eastwards through the snow, aristocrats and landowners, on their way to die labouring in the salt-mines or the taiga. He said it cured him of socialism and he had always owned shares on the stock exchange thereafter if only as a political gesture.

I got on with my work. An old man was wheeled in on a trolley and the nurses dressed him in clean pyjamas and then I examined him with their help. A doctor's note on his clip board should have been very helpful, after all, it was written by a colleague in general practice who knew this man and should be able to describe the problem succinctly. But this one was just a

---

[17] Yes, hospitals were not smoke free, staff and patients alike puffed away everywhere.

scribbled *'Please admit, ? Heart'* or something just as useless and unprofessional. I longed to reply to such letters with just the words *'Heart present'* In place of the usual careful discharge letter with a complete summary of the case notes.

History, examination then perhaps some urgent treatment, oxygen mask, diuretic injection, digoxin to slow the heart rate, aminophylline to help the breathing. Sometimes a friendly word, a clean warm bed and a cup of tea was all that was needed. For some this will have been their last journey and they will go no further. I did my best and took pleasure, even pride, in doing it. I would come back to them in an hour or two when X-rays and an ECG had been done, lab results pinned to the case notes.

The ward staff were good. They got the sterile equipment ready without asking: cannula and trocar for a lumbar puncture; chest drain for the man with a malignant pleural effusion. Are you going to do the bone-marrow this morning? Can you reset Mr So-and-So's drip? I walked round the ward to visit every bed twice a day, the staff-nurse taking notes, I drew blood for the lab, checked the colour of their sputum, changed their antibiotic, examined their chests, estimated their oedema, looked over their charts for fever and urine output. Always a friendly word, even for those in coma, firm words for those not eating or refusing to exercise enough. At some beds we re-examined someone as I worried that we have made the wrong diagnosis and were missing something. Some of the men stopped me with news from home or to ask for a visit from the almoner[18].

I got used to death and dying: in fact, many patients had been admitted to do just that. Others unexpectedly suffered a cardiac arrest. Walton's first intensive care ward would not to open until the following year but we did have a 999 system and a specialist team who would arrive within minutes for patients who suddenly collapsed. Cardiac massage and artificial respiration were relatively new procedures and sometimes, even often, we actually managed to resuscitate someone. I became inured to asking relatives to allow a post-mortem on those who succumbed. We housemen competed to get the highest PM rate for our own ward. If in luck, we were

---

[18] Almoner. What we called social-workers in those days.

asked to sign a form authorising cremation and this brought us a small fee, we called this 'Ash Cash'.

I learnt a lot in the mortuary; it was de-rigeur to attend the post-mortems performed on one's patients. Usually it confirmed your diagnosis, sometimes it did not. Dr Lodzinski was the pathologist and like many of his profession a cheerful man despite dealing with the dead all the time; at least they seldom answered back. He would greet me as 'the doctor that kills his patients by breaking their ribs', a reference to our first meeting when the corpse was of a patient to whom I had given unsuccessful cardiac massage causing, as it often does in the elderly, a few fractures to the rib-cage. One victim was an old man who had died from bronchial carcinoma and whose chest I had drained of a massive effusion more than once. It didn't take Lodzinski long to find we'd made the wrong diagnosis; the old boy didn't have cancer at all but extensive pulmonary tuberculosis. I was mocked again in broad Polish (or perhaps Yiddish) and was left to comfort myself with the thought that in fact my superiors had missed this possibly curable disease in the first place. But that was the point of post-mortems, they sharpened you up by revealing your errors. There's an old joke about doctors burying their mistakes, but in reality we had to know when we erred even if it dented our egos.

Not all the patients were seniors, among the younger ones were a considerable number of people who had attempted suicide and whom the hospital had managed to snatch from death's jaws. One typical evening the casualty nurse called me down to AED, as the huge, brand-new Accident & Emergency Department was called. The previous Casualty was a pokey Victorian slum that had been closed and demolished hardly 12 months before. She wanted me to admit a woman in her twenties who sat on a trolley, quietly weeping, totally exhausted, her hair bedraggled and encrusted with vomit. She had swallowed most of a bottle of Aspirins. Before the days of small blister packs, they sold bottles containing a hundred aspirins; ideal for taking one's own life.

I noted her bandaged wrists for she had also tried to cut herself. The casualty staff had washed out her stomach and I put up an IV line before admitting her to U-ward. That night she was the third admission who had attempted suicide and, to be honest, we juniors, overworked and overtired as we were, had scant respect for 'overdoses'. Their cries for help seemed pathetic to us, why could such people not cope with their lives' problems?

Looking back, I can only feel ashamed at my cold-heartedness. We junior doctors certainly had our own problems with the pressures of our job. We were overtired and were often called to take responsibilities at or beyond the limits of our ability. We were not offered any emotional support or counselling; the mentoring of juniors was unheard of. But what we seemed not to understand was that for all our own difficulties we were really in control of our lives. While we were badly paid, always tired and had little time for leisure, we looked into a rosy future. Many of our patients were not so fortunate; theirs were uncertain futures economically and emotionally; their families were often unsupportive and preoccupied with their own misery; they were poorly educated and many had been abused as children.

That is however what we felt at the time. The sort of girl we preferred was the famous young lady at whose paternity hearing one of our surgical-registrars had to give expert evidence. Her parents made her go to court to ensure that her young boy-friend faced up to his financial responsibilities as the father of the baby that had, incredibly, been conceived while he was a patient in the genito-urinary ward at Walton. He denied nothing she said in evidence and the young couple were obviously in love. He had been admitted after falling off his motorbike at high-speed fracturing his pelvis and several other bones. More seriously the pelvic injury involved his urethra which was torn away from the bladder. This was repaired by the surgeons and a rubber catheter was inserted via his penis into the bladder to support the healing area. Because of his fractures his lower body and legs were partially encased in a plaster of Paris cast and he was unable to move, he lay on his back with one leg supported in the air by a crane while the bones healed. The young woman's evidence, confirmed by the boy-friend, was that when she arrived to see him at visiting time she would draw the curtains around his bed, disrobe, remove his catheter and then climb on top of him. Satisfied, she then reinserted the catheter, drew back the curtains and left. Our surgical-registrar told the court that this was quite impossible; only an expert could have re-inserted the catheter, indeed he himself had had to take the young man to the operating theatre to do this. Nevertheless, the young couple stuck to their version of events. Did they live happily ever after? I don't know, but as usual, love probably found a way.

Accompanied by two swarthy warders a young man, whose leg was enormously swollen, was admitted to one of our side-rooms from the local clink. He was very ill with septicaemia and a high fever caused by an infected wound on his ankle. My opposite number, Joanna was on take and she set about examining him and ordering tests and medication. She set up an IV drip and then insisted that the poor man should no longer be shackled to the bed. For her this was a matter of principal and the handcuffs were duly removed after much warderly tooth-sucking and shaking of heads; Joanna made an impassioned phone call to the prison governor. Joanna was adamant and as usual got her way. At tea-time the warders changed shift and left alone for hardly a minute, Joanna's patient, as sick as he was, jumped out of the fire-escape window onto a neighbouring roof and was away. Liverpudlians were certainly not a passive breed.

One new patient, Mr Smythe, seemed to be in the wrong place, first of all he was very much middle-class with a posh accent. Secondly his abdominal pain was quite obviously due to appendicitis and he should have been admitted directly to the surgical ward and not where he was, in the medical unit. I called the duty house-surgeon, Chris, a tall skinny man with a flowing mop of blond hair. I knew him as a somewhat disgraceful, if amusing character in his own right. An extrovert and definitely a maverick, or, as one might say, a typical surgeon! Chris examined the patient and confirmed my diagnosis and proposed to send 'Mr Smith' to theatre straight away. 'I'm Smythe; not Smith' said the patient in his plummy voice, to which he got the reply 'Smith, Smythe; Shit, Shite; it's all the same to me'. As Chris bounced out of the ward to arrange theatre and anaesthetist, Smythe/Smith just smiled, then winced with pain. I went to write him up for the pre-op medication.

A very old man, Daddy Nic, was admitted. As Dr Nicholson he had been a physician there at Walton Hospital many years earlier; he was now dying from metastatic prostate cancer and was being looked after by Sandy

Skene who had been his junior at some time in the distant past. In pain, he nevertheless refused to have a morphine injection; in fact, he was very upset that it was even suggested. When I checked with Dr Skene, he told me a story that made me think. Many years ago, long before he fell ill, Daddy Nic had extracted a promise from Sandy that if he was seriously ill and when things got really bad Sandy would give him the coup-de-grace and euthanise him. Sandy now regretted this promise, lightly given years ago, as Daddy Nic was now in terror that every offered injection was to be Sandy's making good on the vow.

To me euthanasia does seem to pose more problems than it solves, I cannot shed my almost visceral objection to a doctor deliberately killing his patient. I admit to have very occasionally wished deep-down that a particular patient in extreme suffering would just hurry up and die, but then on further reflection I've usually thought again and somehow found another way to help them that had not at first been obvious. Nevertheless, the answer certainly does not lie, as it does today, in turning a blind eye to suffering patients going off to Switzerland to die while prosecuting those who try and help a much-loved friend or relative to commit suicide at home.

My worry is that if assisted suicide or euthanasia is to be part of medicine there is a great danger that patients will be 'put out of their misery' sooner than is necessary when their caring doctor, understandably deeply involved emotionally with their suffering and feeling perhaps a degree of guilt at his or her own inability to help, may offer euthanasia just as something they can do. To my mind there is an argument for establishing a completely separate euthanasia service which is not part of the NHS and has separate staff who, to put it bluntly, 'swing into action' when the doctor caring for a terminal patient who has had enough and asks that his life be ended. Then patients like Dr Nic would no longer have to worry that they might be killed out of mercy before being ready themselves for such a step. It is certain though, that there is a clear case for those in uncontrollable pain, those whose dignity in life has disappeared and those with frightening paralysis of the muscles of breathing and swallowing being allowed to choose an accelerated death.

A young man in his twenties was rushed into the ward, blue and breathless after a sudden pain in the chest. He had pneumothorax; an air-sac in one of his lungs had burst and air was now leaking into the pleural cavity, a normally closed potential space between the rib cage and the lung itself. As the air gets in and cannot escape, the lung collapses on itself. The signs are obvious, the affected side makes a resonant sound when percussed with a finger on the chest wall, the apex of the heart is shifted, the windpipe is pulled to one side. This man's X-ray film on the right side looked much darker than usual with no lung shadow. After numbing the area with local anaesthetic, I cut through the skin and muscle just below his right collar-bone, and using a metal tube with a sharp pointed insert as thick as my little finger I pushed into the chest cavity. When I removed the sharp trocar and slipped a rubber tube through the cannula, I could hear a rush of air as the pressure inside his thorax was released. He looked better immediately. After cleaning up the operation site I stitched the tube in place and attached its end to large glass jar half-full of sterile saline solution. Air bubbled out from a tube below the water-level and exited from the jar at the top. As he breathed so the bubbles appeared. When they eventually stopped, we would know the ruptured air-sac had closed itself off again. He reclined in bed, anxious but relieved, the jar sat on the floor under his bed, gently bubbling like a hookah. A bit later my bleep went off, calling me back to X ward. The young man had collapsed. He was desperate, blue and gasping. The jar was now up on his bedside table, almost empty. All that water had drained back inside his chest cavity, I put the jar back on the floor and it slowly filled again with water. 'What idiot?.....' 'Calm down doctor, it was the cleaner; she needed to mop around. I've told her and she feels bad enough that she nearly killed the lad'.

The first few years as a junior doctor are famously stressful with lack of sleep, repetitive drudgery and the occasional disaster. There is a constant fear of making a mistake, of causing harm. The frequent feeling of being out of one's depth, the unwillingness to call for help from our equally overworked more senior colleagues. This is a time when long-term personal relationships suffer as well. From happy-go-lucky student to exhausted grumpy houseman is not a change that spouses and girl- or boy-friends find easy to accept. Of course, we all knew to expect this to be so, as we had seen our friends in senior years go through this metamorphosis from student to doctor and, for most, their relationships had survived. For a few it had not; but the transition was not straightforward. In my case I think that my girl-friend, Sue, had expected us to marry after Finals. We did not as I felt certain that we would need to let some time pass and find how we would adjust to the change. As a student I could choose when to study and when to play, now there seemed no time to play or even get enough sleep. I jealously guarded my free time and poor Sue rightly felt neglected. We managed for a while longer but gradually we drifted apart. Looking back, I can see that I must have seemed self-important and selfish with my time and energy, too much of which I put into my work. 'All work and no play' as the saying goes. I don't blame her for leaving me.

There was a rumour in the hospital that we could prescribe a daily bottle of stout for every adult in the ward; apparently the nearby Guinness Brewery generously donated the beer to Walton Hospital's patients and had done so for over 50 years. I checked with the pharmacy and indeed this was so. All the housemen decided to prescribe Guinness as it was bound to help improve their patients' nutrition, wouldn't it? As we had hoped, most patients refused it, but it continued to arrive by the crateful in the ward. Our charge-nurse grumbled about it getting underfoot. Surprisingly, somehow, it ended up in the doctors' sitting room. There was too much even for us to get through. We stopped prescribing it. The charge-nurse said that this happened every year.

Christmas approached, blobs of cottonwool 'snow' were glued to glass partitions, paper streamers looped across the ward, the hospital radio played Jingle Bells interminably, mistletoe hung enticingly in the freezing sluice. One late December evening the local church choir arrived at the ward doors and struck up with 'God rest ye merry gentlemen' and then moved gradually down the ward from bed to bed as they sang their way through the usual repertoire. As they sang, delighting everyone, one of our old men collapsed in a lavatory cubicle and we rushed to the sluice-room to save him. The cubicle door was jammed as he had fallen against it. I climbed over and opened the door from the inside as the emergency team arrived. All the usual activity; intubation, oxygen, cardiac massage. The ECG showed ventricular fibrillation so we gave him several shocks with the cardioverter. Despite our efforts he died and we called the mortuary. The porters arrived, picked him up and wheeled him away and out of X ward in their special trolley. The choir, completely unaware of the drama occurring a few metres away, was nearing the end of their performance and as the singers politely moved aside for the deceased in his closed chariot, they harmonised away without dropping a note. Merry Christmas.

On Christmas day there was a tradition on the wards that at lunchtime the houseman carved the turkey, the consultants waited on the patients and we all celebrated the holiday together. Not that year though; a new catering officer declared this to be wasteful and so patients were to be served pre-plated lunches, the turkeys were to be carved in the kitchen, the doctors were no longer required. There was even some question about whether free meals should be served at all to the staff. Ebenezer Scrooge stalked the wards.

In this spirit of economy, the doctors agreed that the medical dining-room could be closed on the 25th and those on duty would join the nursing staff in their canteen for meals. We arrived at the canteen to get the Joseph and Mary in Bethlehem treatment. Nobody warned them about this, they had only prepared enough meals for the nurses. The doctors were none of their business. Muttering insurrection, we retreated and I was volunteered to ring the hospital secretary at home, no doubt at that moment himself sitting down to a festive family meal. His response was immediate, I was amazed, what a good man! Within an hour he had arrived in the hospital with his wife and they re-opened the kitchens and his staff prepared us something to eat, not a turkey, but it was a good hot meal. We were assured that heads would roll!

The New Year passed and soon it was time for me, sadly, to leave Walton and move to the General Hospital in nearby Ormskirk for 6 months as house surgeon.

Ormskirk is a pleasant market town in the Lancashire countryside, nearby was the new-town of Skelmersdale; Skem is where many Liverpudlians had been rehoused after slum clearances in the city. Despite its bucolic street names, Skem was not a pretty place. Our patients were a mix of both persuasions; rehoused Scousers with their broad Merseyside speech and the rural Lancastrians speaking in Northern tones.

I was to be the houseman on the surgical ward, Mr Marsden whom I'd worked with before in St. Helens, and Mr Wright were my two bosses. Much smaller than Walton Hospital, Ormskirk stood in leafy grounds and the doctors' residence was a detached house nestled in a garden behind the huge block of the mental wards. The hospital, apart from the Victorian era lunatic-asylum block, was quite small which meant it was a bit quiet and, apart from the psychiatrists, there were few specialists; only general surgeons and physicians. No paediatrician, no cardiologist, no orthopods. Specialist problems had to be ambulanced into the city half an hour away. It was seeing how ineptly one very sick child was managed by these generalists that set me on the course to become a paediatrician; I will come to that.

My life in Ormskirk was a bit humdrum; my two surgeons did their humdrum lists. Hernias, gallbladders, gastrectomies; removal of general lumps and bumps and the stripping of varicose veins; old men lost their prostates. I assisted at four operating lists each week. My job was to get the patients ready for their operations, ensuring they were fit enough for surgery and that the diagnosis had not changed and had even been right in the first place. I sorted out diabetics who would not be able to eat for a day or two, treated high blood pressure, made sure the paperwork was all done, especially the important consent form. Then with the nurses helping and checking, we marked which side was to be operated upon with a felt-tip pen; it would not do to remove the wrong leg from the wrong patient and it looked bad when you had to wake them up to ask when you were not sure.

After surgery, as the patients returned from theatre, I made sure IVs were running and analgesia was written up. I checked the operation site was not bleeding, sent blood off to the lab. and then rushed back to theatre to see if they would let me do something at the end of list, perhaps some minor procedure. A sebaceous cyst or a skin cancer, perhaps stitch up after something more major. My left-handedness was a handicap as I needed to stand on the 'wrong' side of the patient and theatre sisters did not like the inconvenience this caused. I am not particularly dextrous and had no ambition to become a surgeon but I did get an inordinate satisfaction from doing minor procedures. Eventually before the end of my six months stint, I had done a few appendicectomies on my own as well as a simple hernia and several circumcisions. Very easy stuff, but as I say, satisfying.

The psychiatrists asked Mr Marsden to see a patient in their part of the hospital, an intellectually-challenged youth with a mania for swallowing any metal object he came across. On his X-ray we could identify bath plugs on chains, a ring of keys, many bed-frame links and springs, razor blades and teaspoons and a small salt-cruet. Nowadays a flexible gastroscope would be used to get this lot out but in 1969 the Japanese had not yet put the finishing touches to inventing endoscopes, so our surgeons had to rely on a surgical approach. Mr Marsden opened the youth's belly and removed several kidney-dishfuls of metal items; some going to rust and some still bright and shiny. Before closing-up, an X-ray showed we had removed everything and the lad was returned to the surgical ward where we had prepared a side room completely devoid of any small metallic objects, or so we thought. Within an hour the staff-nurse could not find the little watch she wore pinned to her apron and the drug-trolley keys were missing. We planned to get the lad back to the mental-unit as soon as possible but by the time he left us a final X-ray revealed many items apart from the keys and the watch, including several shillings in copper and silver coins, a Parker fountain-pen cap and a disposable scalpel blade.

The sick child whose inept treatment had such an effect on me was a little boy about 6 or 7 weeks old with severe vomiting. There had been delays before getting him to the hospital, and when he arrived he was dehydrated and in a state of collapse. An attempt was made to get him to drink from a bottle of water but this he promptly vomited back in a great arc of a metre or more across the room; the classic projectile vomit that is virtually diagnostic of pyloric stenosis. I could in fact feel the lump of the swollen pylorus that was blocking his stomach when I laid my fingers gently on his tummy.

There is only one way to deal with this problem; after rehydrating him you take the baby to theatre and incise the pyloric swelling under general anaesthesia. This produces a permanent and immediate cure. I rang Mr Wright with the news and he appeared bringing the duty anaesthetist. We needed to put up an IV said the gas-man and we searched the tiny body for a vein. The baby was actually going into shock from lack of fluid and would almost certainly have abnormal blood electrolytes as vomiting quickly depletes essential Potassium reserves. Even had we been able to find a vein nobody knew what fluids to give, nor how much was needed. Mr Wright decided to cut down through the skin of the leg and find a vein to cannulate; I was sent to the library to research what fluids to give a baby. After a couple of hours, the patient was looking better and had passed some urine. Then the anaesthetist was not sure how to give an anaesthetic to so small a child; again, Mr Wright was resourceful. His plan was to let the child suck on a dummy dipped in Nepenthe, an old-fashioned oral morphine mixture that completely zonked the child. Then using local anaesthetic, he opened the skin and abdominal wall muscles and located the swollen pylorus which was a glistening white round lump. It looked rather like a pickled cocktail-onion under the theatre lights. He incised it neatly and we watched it quickly retract on itself, releasing the blockage. The baby's tummy was stitched up and he was returned to the ward where I spent the next hours watching him suck small amounts of sugar water from a sponge. In 72 hours, he was lusty and ready for home; a miracle indeed, a lucky miracle to have survived despite our appalling ignorance.

This episode woke me up, how could I call myself a physician if I could not treat such a common condition as dehydration in a young child? I realised that I was totally ignorant of how to resuscitate a dehydrated baby, how to set an IV in a tiny vein, what fluids to use and so on. How had I spent 5 years at medical school without being taught this? That little boy could so easily have died in our hands.

That week I applied to the two children's hospitals in Liverpool for my next job and the course of my life was reset again. The surgical job dragged on uneventfully through the summer until in September I was promoted to SHO (Senior House Officer) with a useful rise in my salary; and I moved to the Royal Liverpool Children's Hospital on Myrtle Street.

CHAPTER 9. MYRTLE STREET.

There were two hospitals in the city for children at the time; the Royal Liverpool Children's' (RLCH) on Myrtle Street near the middle of the town centre and the university campus; and Alder Hey out in West Derby near Knotty Ash, famous at the time as the home of the comedian Ken Dodd. Alder Hey was a huge place spread out on a large site. RLCH was smaller, more compact, it had a Casualty Ward on the street level, the wards, with huge bay-windows to let in the sunlight, were at the front of the building and the operating theatres upstairs at the back. Liverpool's main paediatric heart-surgery unit was at RLCH.

A few months before my arrival at Myrtle Street a newly appointed matron had decreed that parents could be with their children in the wards at all times throughout the day and night. This revolutionary idea so shocked some of the senior nurses, appalled that mothers and fathers would be interfering and getting in the way, that they resigned in protest. The move had however proved a great success and now, only a few months after the great change, the nurses would be rather sniffy when a parent did not accompany their child when admitted to the wards; 'who do they think we are? We don't have time for feeding and changing these children' they would complain.

I settled into the doctors' residence across the street from Casualty and within minutes was called to my new ward. My new boss was the celebrated paediatric-surgeon Mr P.P.Rickham. He was now approaching retirement age and was incensed that the University had never awarded him the professorship he felt was his due. Back in Germany, where he originated before coming to England as a child, all his contemporaries called themselves 'Professor', as is the habit in German speaking places. His achievements had certainly earned him the award of a personal chair, but he must have upset the wrong person sometime in the past and Liverpool was firm in its refusal of the title. In post-war Liverpool PPR had been a pioneer in children's surgery and had developed techniques for dealing with many congenital conditions including Spina Bifida, where he had shown that early closure of the defect in the back improved outcomes. His operation to reduce the commonly associated hydrocephalus using a valve to drain the excess cerebrospinal fluid from the brain is still used today. He was justified in his expectation of this recognition for his work and, frustrated in this, he was soon to quit Liverpool for Zurich.

For three months I was to be his House Surgeon, I set up IVs, took blood, prepared kids to go for surgery and provided post-op care. I assisted the great man in the operating theatre and sat at his feet, metaphorically, in out-patient clinic. The senior-registrar on the unit was a hugely tall Nigerian called Musa who shared the clinics with PPR. One afternoon a small boy of African descent was sent in by his family doctor needing circumcision for a tight foreskin; his phimosis had already caused several painful bouts of infection. Mr. Rickham agreed with the GP and was arranging admission to the ward when the boy's mother had hysterics and flatly refused to have her son treated. The family was from the local Nigerian community and PPR wondered if this lady, though smartly dressed, obviously well-educated and speaking good English, misunderstood what was intended. Musa was called in to have a chat with her. Musa towered above the little woman and absolutely roared at her in his deep bass voice; she dropped to her knees, grovelling at his feet and touching his shoes, she looked a bit shaky when persuaded to get up and immediately agreed to the surgery. I don't know what was said, though it didn't sound to me like Musa was actually explaining anything about the operation. Obviously, our Musa had some influence in West African society.

I spent a year at Myrtle Street. After surgery it was cardiology under the not so tender eye of 'Shack', Jean Shackleton, a terrifying female consultant. Then I went to the casualty department and finally general paediatrics. I had learned the necessary skills, putting up drips in tiny babies, calculating fluid regimes, treating diabetics and asthmatics. I could now diagnose congenital abnormalities of the heart, intersexes, kids with ear infections, hernias, broken bones, meningitis. It was busy and demanding work and I loved it.

The patients' waiting room in casualty was dominated by an enormous teddy-bear; about 8 feet high and almost as wide, it had been donated by the Rotary Club decades previously. One night it was stolen. While we were all there working away, someone nicked it; though how they got it out through the corridors and narrow exit doors I can't imagine. Though the police were called, the casualty sister was delighted as it freed up several extra waiting-room seats for patients.

One of the local families, blessed with numerous children, made frequent visits to Casualty. Mrs Quinn was always in a muddle with one or other of her offspring. Little Jimmy Quinn was about 3 years old and often had to be admitted with the 'bronnicles' as she called his asthma. After a few days of treatment in our nice warm ward he went home to his cold and damp house where his mother got in a muddle with his medicine again, and back he would come, usually at around 2 am.

Young Myrtle Street Denizens

I arranged on one of my afternoons-off to accompany the almoner, as social workers were then called, to chez-Quinn about 500 yards from the hospital in Liverpool 8 near the Rialto which was still a dance-hall at the time. Their old building was damp and depressing, the Quinns lived in two unheated rooms on the first floor; both of which contained several beds and a TV set. We managed to get the family rehoused in a 4-bedroom flat in a modern building a few streets away with central heating, bathroom and hot water. A bit later I went to visit the Quinns in their new accommodation and found them all crammed into one room, the others all left empty and unused. Mrs Quinn said she didn't like the central-heating and had bought a paraffin heater round which they all gathered to sleep. The ceiling dripped with condensation; Jimmy sat wheezing in front of the TV set.

Although asthma had many of the features of an allergic disease no allergen was consistently found to be involved in its causation. One of the senior child-psychiatrists proposed a theory that postulated a psychological origin for the disease. After all, nobody had a better idea at the time and the parents of a child with severe asthma were often at their wits' end. Professor Rose's theory went that there was particular family dynamic when a child had asthma; the child could not run or play football and was a disappointment to his father who then, subconsciously, rejected him and thought him a weed. The mother overcompensated for this and thus her mothering became 'smothering'. So, the poor child developed psychosomatic symptoms and wheezed. That was the theory anyway. Professor Rose further argued that the solution to this problem should be 'parentectomy'. The child had to be separated from his parents who would then be subjected to re-education in correct parenting. Wheezy children were then snatched from their loving families and sent to a disused and freezing-cold TB-sanatorium high in the Pennines for weeks on end, where indeed their chests usually improved. We clinicians thought this was cruel nonsense and tried to hide our little asthmatics from Professor Rose when he prowled round looking for victims.

A year or so later research in Leeds revealed that allergy to the house-dust-mite was the major trigger for asthma attacks, this minute arachnid was hitherto unknown to us doctors. Of course, the truth was that the sanatorium in Buxton was kept so clean and so well ventilated that the house-dust-mites couldn't survive there and so naturally most kids stopped wheezing. It's an old saying among researchers; 'Association does not prove Causation'. That was the end of parentectomy.

A bleep in my pocket. Paediatric SHO at Myrtle Street.

A problem then emerged in the surgical departments; an unusually large number of patients developed post-operative infections and so the main theatres had to be shut for a few weeks for deep-cleaning and maintenance. One evening I was enjoying a nice soak in a hot bath in the doctors' residence when some bird's-feathers spurted out from the tap and floated on the surface. I put them in an envelope and later showed them to the hospital secretary. On investigation they found that the lid on the main water-tank in the hospital roof was damaged and indeed the remains of a dead bird could be seen in its depths. Nobody said anything but, just perhaps, this was connected with the infections as surgeons scrubbed up in contaminated water before operations.

The canteen food was pretty awful and eventually I went to complain to the catering staff about it, after all we were confined to the hospital for days on end and this was long before the days of Uber-Eats delivering Pizzas or Chinese take-aways to the door. My boss at the time was Professor John Hay who demanded that I come to his office and talk to him about this lese-majeste. That afternoon I was helping his wife Netta in the dermatology clinic. She had heard that I was to be carpeted by the great man and she told me a tale about when he was a houseman at this same hospital. He had been so annoyed himself at the bad food that he had carved the date in the rind of a piece of dried-out cheese and it had still been there on the cheese-board weeks later. At our meeting he was all smiles and said that Netta had had a go at him; he had now been to the catering officer himself and hoped things would improve. They didn't, but this was a lost cause; all hospital food was (and is) awful, everywhere. Though I fondly recall one small exception to this; the chef at Mill Road Maternity Hospital was mightily celebrated for his delightful Crème Caramel puddings. Alas he died young a year or two after this from cancer.

In the canteen I often shared a table with Mohammed Zubaire. Zube was the senior registrar in cardiac-surgery, he was immensely competent and always appeared immaculately shaved and dressed, even in the small hours of the morning. He and the senior anaesthetist Gordon Jackson Rees were the backbone of the hospital. If you had a really sick child on your hands 'Jacko' seemed to materialise at your side, day or night, to help. He was another pioneer in the post-war years and I understand that anaesthetic equipment he originated then is still routinely used today.

Zube was what was then called 'time expired'. Fully trained and absolutely competent to be a consultant surgeon he had been persistently passed over when promotions came up. He was from Pakistan and racism might have been involved though other foreign graduates had been advanced. One evening he was particularly upset after he had been to fetch a patient's notes from his boss's office. While there he had seen his own personnel file in which his boss, whom Zube had regarded as a friend and a supporter, had written a very cold testimonial to accompany Zube's latest job application. After I left the hospital, I kept in touch with Zube and he did get his consultant job soon after. This time he had asked someone else to provide a reference.

One night I admitted a young girl with a high fever and slight jaundice, the family doctor's note suggested viral hepatitis. However, she actually had Weil's Disease (leptospirosis); her eyes were inflamed and there was blood in her urine. 'How had she caught this?' demanded her father. I told him that it was spread by animal urine and rats were usually to blame but there was a variant caught from dogs. He replied that there were no rats in his house but 'the neighbour's bloody dog was always pissing on our doorstep'.

After a few days on antibiotics the girl was ready for home but the parents didn't show up to collect her. They had been arrested for a violent assault on their neighbour after first killing the dog with an axe.

These were not the only parents who were perhaps excessively passionate in the defence of their offspring. One night I had to call the duty surgical registrar, George Roberts, to come in to the hospital from home, to see a schoolboy with appendicitis whom I was looking after in Casualty. George hurried in followed by a young policeman who was trying to detain him. George explained the urgency of the situation but the policeman felt that breaking the speed-limit while driving to the hospital in his fancy Ford Capri GT was nevertheless unwarranted. George broke away heading for the casualty-ward and again the officer held him back, at this point the patient's dad gave the policeman a good biff on the jaw. We managed to delay his leaving for the police-station in handcuffs just long enough to get his signature on a consent-form. George removed an inflamed appendix and the boy soon went back home, perhaps before his father did.

I was considering taking the DCH[19] examination but Professor Hay dissuaded me saying it was not really intended for career paediatricians, more a qualification for generalists working with youngsters in general practice or the schools' health service. It concentrated too much on developmental problems rather that clinical medicine. One feature it seemed to emphasise was the recognition of 'patterns or human

---

[19] Diploma of Child Health.

malformation'. These syndromes[20] were recognisable from the physical build and appearance of children, particularly their faces. The obvious example that everyone knows is Downs Syndrome. Children with Downs are instantly recognised by their facial characteristics and the flattened back of their heads, but there are many other such groups of features such as Treacher-Collins and Foetal Alcohol Syndromes. In those days learning the syndromes was necessary to pass the DCH and the Prof rightly thought this emphasis distasteful; although specialists had to be able to diagnose them on sight so as to prevent unnecessary investigation of the cause of developmental delay in children who showed the features of one or other of the syndromes. One warning was always to be heeded, the doctor must first take a good look at both parents before plumping for one of these diagnoses as some kids just looked like their dad and there was nothing wrong at all. I was advised to concentrate on passing the Membership exams and forget DCH.

One nice thing about paediatrics was that we saw few if any attempted suicides though accidental drug overdoses were common. Two-year-olds getting into their mother's handbag and eating a handful of prettily coloured pills was very common; contraceptives and vitamins would cause little harm, while anti-depressants, sleeping pills and paracetamol could be lethal. Locked medicine cabinets were not common amongst the population in Liverpool 8 and when granny came to stay the row of pill bottles by her bed could be irresistible to a pre-schooler. Planning for better drugs safety was in the air but analgesics were still sold in bottles of a hundred pills. The Myrtle Street hospital pharmacist was looking for childproof containers to use in our dispensary and one morning asked the children in our ward to test a new tablet-bottle he was considering. A few Smarties were put into the specially capped containers and handed out to the little terrors under our care. It seemed the design was a success as they tried and tried and failed to open the lock. Then one child simply bit off the end of his plastic bottle to get at the chocolates, then all the others copied him; another failure.

---

[20] Syndrome: A recognised disorder defined by a group of symptoms and signs.

The usual management of children who had swallowed medicines or other noxious things, was to give a few spoonsful of a syrup containing ipecacuanha. Within a few minutes this provoked vomiting. Much kinder than using a tube, it was in fact more effective in emptying a child's stomach. We usually kept the child overnight for observation but mainly so sister could give the parents a piece of her mind! When Gail, age three, was admitted her mother swore that there were no pills in her house that the child could possibly have swallowed. Gail was manic, rolling, jumping and jerking in her cot, banging her head, unable to keep still, her skin was flushed and her pupils dilated, heart rate rapid. Despite her mother's denials I was sure this was atropine poisoning. A derivative of atropine is hyoscine, a pill I knew well as it was used by yachtsmen as seasickness prevention and was sold as Kwells. 'Oh, Kwells' says Mum 'but they aren't medicine, I just take them for my bus ride to work'. Gail had eaten all the contents of the tube of tiny pink pills and the empty container lay in her handbag when we looked.

CHAPTER 10. MOSSLEY HILL.

As I approached the end of my year of paediatrics I considered how to progress. I had originally intended to spend six months learning the skills of caring for sick children before returning to general medicine. Though I had no goal at the time beyond a desire to gain the MRCP[21] and become a physician; to me the highest calling in life. Instead, I signed up for a second six-month stint at Myrtle Street. Why? Because the paediatrician's approach to a child and his disease was intensely satisfying. It seemed that in Child-Health you had to care for the complete person; you had to consider the patient's place within a family, how his illness affected the other siblings and his parents and their livelihood. Further your patient had to be seen as a developing individual, physically, mentally and intellectually. An acute condition like appendicitis or pneumonia might have only a brief effect on the child as a whole but chronic disease such as asthma or cerebral palsy changed his life and that of the family. I started to understand how attention to the whole person in his environment was the key to good health-care. I felt that my previous approach in adult medicine was incomplete; There I had had no time to do more than sort out the disease, whereas sorting out a patient's other problems, even when they had a bearing on his recovery, was someone else's concern; almoner, relatives, vicar or whoever. I had healed him and my job was done. In paediatrics there was another layer.

I was impressed by the writings of John Apley and particularly of Ronald Illingworth[22] who wrote about children as complete persons and how childhood experience affected their adult lives for better or worse giving examples of famous men and women or extracts from the writing - diaries, memoirs, biographies - of less celebrated individuals. To help young people to overcome or even benefit from their handicap or experience of illness and enable them to become better individuals seemed the worthiest of callings. While I well understood that as a doctor my main duty was to bring

---

[21] MRCP. Member of the Royal College of Physicians. The postgraduate qualification for a specialist in internal medicine.
[22] Professor of Paediatrics at Sheffield University who wrote extensively, including *The Normal Child* and *Lessons from Childhood*.

my scientific knowledge to bear, I came to realise there was more to it than that.

Nowadays paediatricians in training remain in child-health and work through the various specialties, rotating through oncology, neonates, developmental, endocrinology and so on, then they take the all-important MRCPaed[23] exams. In the 1960s the Paediatric Royal College was yet to be founded and we had to take the adult MRCP exams. So back to adult medicine I had to go with the intention of returning to paediatrics in a few years' time once I had passed the adult exam.

Mossley Hill Hospital was just round the corner and a few hundred metres from my home in Number 4, so I could walk to work. Set in a smart residential tree-lined street, it had once been a fine old family-house and it had been turned into a hospital for returning soldiers in the 1940s, catering especially for prisoners of war. When I arrived in 1970 it was still something of a quiet backwater where only selected patients were admitted, mainly arriving via the outpatient clinics rather the emergency services; in other words, 'cold-cases' were in the majority. Some came for investigation, others on long-term treatment were being repeatedly readmitted for follow-up courses of therapy. This made for an interesting mix of people, often with unusual conditions, rather than the usual frantic rush of the acute hospital. I now had time to study; to read the textbooks, to get to grips with recent advances by reading the journals and even to write up some cases for publication. Over the previous two years I had been learning the nuts and bolts of hands-on practical doctoring; now my period at Mossley Hill was going to help me develop a firm base of knowledge for the next phase of my career. I was to be on call every third night and weekend; on most days I managed to leave the wards before 6pm; it was just like a normal life.

When they returned from the prison camps, especially those in Asia, many soldiers were suffering from malnutrition, chronic malaria, tropical infestations and especially post-traumatic stress disorder (PTSD), though that name was yet to be coined. Even 25 years after their release from the Japanese, ex-prisoners would come into our wards for a few weeks to have

---

[23] MRCPaed, Member of the Royal College of Paediatricians.

their symptoms treated. Chronic malaria was the usual excuse though they had probably been free of plasmodia for many years; their weakness, shivering attacks, weight loss and so on were probably not due to persistent Asian parasites but to PTSD. Indeed, we did see the occasional patient with persistent infections especially those due to amoebae. All these men were heroes who had seen and borne terrible things and naturally they were welcomed back every year or two for as long as they needed. They formed an interesting but dwindling minority of our patient population; some getting over their stress, others passing on.

One of our consultants, Dr Epstein, was a very up-to-date cardiologist with a research interest in coronary thrombosis and myocardial infarction (MI) which was much more frequent in those days when everybody smoked and people's diet and housing were much worse than nowadays. Most of Dr Epstein's beds were occupied by patients with uncomplicated MI, the treatment for which at the time was to stay at home in bed for three weeks of immobility. The local GPs sent them to us if home-care wasn't feasible and we put them in one of our beds. Usually, they gave us little trouble. Central venous lines were all the rage at the time and we became adept at inserting them; according to the then current research protocol some received blood thinners while others didn't; I can't recall which had the better outcome but most went home after their three weeks with advice on avoidance of stress. Stress was regarded as the main cause of MI and we were only beginning to realise that cigarettes actually caused many more diseases than just lung cancer.

Among our patients with myocardial infarction was an American visitor to Liverpool, who had suffered chest pain while staying at the then very smart Adelphi Hotel on Lime Street. He spent the regulation three weeks in our little hospital being subjected to our clinical experiments. On the last day his family collected him in a taxi and he asked for the bill. He could not believe the generosity of the NHS in allowing him entirely free treatment. I left him packing his suitcase and still trying to press money on members of staff.

Mr Harris had been in the ward several times before and his GP asked us to readmit him as his chronic chest disease had deteriorated. Mr Harris had severe emphysema and was classified as a 'blue bloater', as opposed to those in the earlier phases of the disease like Dr Alstead who are called 'pink puffers'. Mr Harris lay there in bed, unmoving except for an occasional deep gasp for air. He was unconscious for most of the time, his face bloated, his skin and lips blue, he was fed via a tube through his nose into the stomach, urine drained into a bag by the bed-rail, an IV-line dripped medications and antibiotics, an oxygen mask covered his face. A physiotherapist came every few hours and pummelled his chest. He was certainly going to die; this was his last innings. His family arrived from New York to make their last farewells.

The Harris family were gathered round their dying patriarch's bed one morning where he lay immobile and unknowing, taking occasional and irregular gasps of oxygen. 'Won't he ever wake up doc? We've come all this way to see him and we want to say goodbye to him'. I thought not, and told them so, but later one of the older ward-sisters asked if I'd thought of using Vandid. This so-called respiratory stimulant was very much out of fashion but some journal articles still recommended it.

Warning the relatives of its possible ill or even fatal effects I proposed trying a single dose and so slowly injected Vandid into the IV line. The result was dramatic. Mr Harris began to move, then struggle. He clawed at his nose and face and he opened his bleary bloodshot eyes to look round at his family. 'F***ing Hell', he grunted and collapsed back into his pillows, dead. I was horrified but the family were delighted; 'Oh thank you doc, he saw us and could say something before he died'. That was the only time I ever saw Vandid used.

Bridget was one of our senior sisters, a strongly republican Irish lady from the deep west of her country. Returning from a family visit she brought me a present; a John Powers whiskey bottle containing an opaque amber liquid. This was 'the real Potheen' she told me, distilled in her own village. I took it home and put it on the kitchen shelf and forgot about it. A week or two later I looked at the bottle and saw that the liquid had cleared and had formed a deposit of white crystals filling about a third of its volume. A friend in the chemistry faculty took it away and found its lead content was in the lethal range. Did Bridget imagine I was a secret supporter of the much-hated protestant cleric Ian Paisley?

Now that I was living a more normal life I made two big changes. First, I gave up my old room in Number 4 and moved into a 3 bedroom-flat at the much more civilised House Number 1, Mossley Hill Drive. Charles and I took over the ground floor garden flat. It had a patch of grass which we assiduously kept mown. Our cleaner from Number 4 was recruited as charwoman. Charles' mother, who knew about such things, advised us to be sure she was a 'good scrubber'. Josie kept the tiled floors spotless, mainly by spreading the dirt a foot or two up the walls. I was very happy there for the next few years. One of the other residents living, in a large flat that included the library of the old house, was the journalist Gillian Reynolds who gave memorable Pimm's parties on her lawn on summer evenings. Upstairs above us in the mews flat was the venerably ancient Miss Clarke with her cat which was famous for being unable to eat anything but fresh prawns poached gently in milk.

Sailing *Galadriel*.

I decided to take up my old sport again and bought a new racing dinghy, a GP14 to be called *Galadriel* and built to my specifications. Charles and I successfully campaigned her round the North West on my weekends off, winning several trophies and getting me physically fit again. All that hospital food and lack of proper exercise had made me put on a lot of weight. At the end of the season Charles announced he had bought a terraced house in Gateacre and he moved out of the Mossley Hill garden flat. Despite Charles' defection I was beginning to relax and enjoy myself; the really hard times were over and I could begin to think of life outside medicine, I had a bit more money to spend and I was over losing my girlfriend Sue. There was still the MRCP exam on the horizon but I was well into my studies for that. Life was looking good especially when I met a medical technician from the Chest Department's laboratory called Joan.

A medical technician called Joan.

CHAPTER 11. WALTON AGAIN.

I was now moving up the ladder as I was accepted onto the three years specialist training-rotation and was promoted to Registrar status. I moved back to my old hospital, Walton, for the first of those years. Sadly, my old boss, Sandy Skene, had died the previous year and Bert Alstead had retired. A new management team was giving the place a shake-up. A young consultant, Mike McCauley, had set up the intensive care unit and the neurosurgeons had, in effect, a brand-new hospital of their own within the main hospital. I was to work for CJ Williams. When I heard this my heart fell. CJ had a reputation for being pernickety and never allowing his juniors to get on with their job by constantly interfering. He was a leading light in diabetic management so D and E wards were full of diabetics being stabilised as well as the general run of medical cases; this is a fiddly business of trial and error. Not only did we have to adjust the total dose of insulin but also select different preparations with various lengths of action. Several times a day we checked the blood sugars. Nowadays the patient can do this himself, pricking a finger and using a meter that will fit in a pocket or a purse; in those days blood was drawn from a vein and sent off to the lab, the result delayed for an hour or two. CJ was eternally asking for more frequent testing and then changing doses of insulin by tiny percentages that could have no clinical effect. This threatened to drive me crazy so I decided to go one better than him. The lab staff and the nurses though I had gone mad but I managed to wear CJ down. I could almost read his mind and whatever he asked, I had already done it, more and more tests, all times of the night and day, tiny changes in doses. The patients must have felt like pin-cushions. CJ cracked first and after that he left me alone. We got along very well and became friends working well as a team. Perhaps our very sensible ward-sister had had a firm word with him.

While X ward had been long and narrow D was a typical Nightingale ward with beds down the middle too and there were also smaller side-rooms for patients who might be infectious or need some form of special attention. One afternoon we admitted a retired army major from

Southport[24] with a heart attack. He had an unstable rhythm and was quite unwell at first and I refused his rather bossy wife's demand that he should occupy one of the 'private' rooms as appropriate to his social status. I wanted him where he could be seen should he collapse. After all, the probably apocryphal rumour was that Dr Skene, after his heart attack, had been nursed in a side-room and nobody had noticed when he suffered a fatal cardiac arrest.

The major would sit in his bed and order the fitter men around to help the ward staff; serving food, fetching and carrying, distributing the newspapers or the mail. When he was better, I suggested we accede to his wife's request and move him to a vacant side room, but he refused, preferring to 'stay with the men'. We were all quite sad, especially 'his men' when his missus took him back to Southport after the regulation three weeks in bed.

As a registrar, now I had several juniors working alongside me and I had to learn how to help them, to teach them and when I delegated duties to find how much I could trust them. We were absolutely reliant on the housemen (male and female) who did all the drudge work. If they were absent for any reason an enormous load fell on the rest of the juniors. Our houseman, Paul, was a bit of a Jock who would ultimately and not unexpectedly, become an orthopaedic surgeon. He had only been with us for a week or two when he managed to get his jaw broken while playing in a hockey match. He bravely came back to the ward within a week with his jaws wired shut and we had to feed him mush down a tube pushed through a gap between his teeth. His speech was incomprehensible and so his history taking was unreliable and there was no point in his using the telephone

One of our SHOs was a bit thick to say the least, but extremely willing; he was desperate to please and hated calling me to check his management. One late night he phoned to say he had admitted a man with a stroke but I was not to worry as he had everything in hand. His diagnosis was of a haemorrhage in part of the brain called the pons; this was suggested by the

---

[24] Southport. Nearby seaside town where the middle classes liked to retire and play golf.

83

patient's pinpoint pupils and high fever. The man was deeply unconscious. There was no history beyond that he had been found by a cleaner in an empty railway carriage at Lime Street station and his suitcase and wallet were missing and had probably been stolen. I was exhausted and noting it was 2 am was determined to go back to sleep in the little bedroom we used when on night duty. But I knew in my heart that I could not trust this SHO's opinions. If the man had had a pontine haemorrhage, recovery was unlikely, so what was I expecting to achieve?

I put on my white coat again and walked over to the wards. The patient was lying in bed, unconscious, his temperature was 39 degrees and it was all exactly as the SHO had said; except, strangely, he was very tanned. In the UK of 1971, nobody had a suntan during the winter. I was suddenly struck by the truth and I knew, with no more evidence than this, that he had cerebral malaria. An hour later the on-duty lab-technician phoned confirming the presence of falciparum malaria plasmodia in his blood-smear. I called the School of Tropical Medicine and the duty doctor agreed with my choice of chloroquine as treatment and said they would come over and advise. When the tropical specialist came over a few hours later he recognised our man as a colleague doing malaria research who had just returned from abroad. In a few days the researcher had regained consciousness and we shipped him off to the Royal Infirmary's tropical disease unit. I could easily have stayed in bed that night and he would certainly have died, probably still thought to have suffered a stroke. It was a lesson; there is little point in diagnosing an untreatable disease until you have excluded every other possibility.

The cardiologist Mike McCauley was interested in the use of pacemakers to treat heart-block during the acute phase of heart attacks. He took each of us registrars in turn and trained us to insert pacing wires into the heart through a blood vessel in the neck. As is usual in medicine one sees a procedure once, then you do it once under supervision and then you start teaching it to others. (See one; do one; teach one). By the time I left Walton I had placed several wires and was passing on the skill to my colleagues.

One skill nobody in the hospital seemed to have however was taking a sample of CSF[25] from the fourth ventricle in the base of the brain. CJ had admitted a man who had been born with spina bifida, a defect where the lower end of the spinal cord is exposed causing damage to the nerves. At birth the defect is usually surgically repaired as soon as possible to prevent further disability. A large area of this patient's skin around the bottom of the spine was infected and he had a fever with early signs of meningitis. We urgently needed to take a sample of his CSF. This is usually done by lumbar puncture but the infected and deformed site in his lower back precluded this. CJ said 'Get someone to tap the fourth ventricle'. I asked in the neurology department with no success and came to the conclusion that I'd have to do it myself. A nice little book with a lurid pink cover, 'Pye's Surgical Handicraft', that I found in the library made it sound easy; 'with the patient sitting, arms resting on a table, neck fully flexed etc.' We shaved the back of his neck, numbed the area with Lignocaine and then, using an especially long lumbar-puncture needle with its sealing trocar in place, pierced through the skin. I aimed for the occipital bone, and gently bounced the needle tip bit by bit downwards over the bone until it entered the foramen magnum, where the spinal cord emerges from the skull. Then changing the angle slightly upwards to point straight at the root of the nose I advanced the needle 2.5cm as directed by Dr Pye until I felt a slight 'give' as it pierced the membrane and went into the ventricle. I removed the trocar and CSF dripped from the needle into a specimen bottle. The fluid looked 'as clear as London gin'; just as it should; and soon the lab confirmed there was no evidence of infection in the brain.

Apparently, this was a sensation, it seemed that no one in the hospital had ever seen it done before, so D ward had been full of spectators. I've never done it again and have never met anyone who has seen it done since either. The patient's CSF was healthy and he soon recovered and he never asked why there were so many observers on his big day.

Once a week each of us medical registrars was 'on' for the whole hospital. Non-medical wards needing a physician's opinion on a patient would bleep us. I came to visit parts of Walton I hadn't suspected existed.

---

[25] CSF, Cerebro-Spinal Fluid. Liquid that bathes the brain and spinal cord.

For example, the psychiatrists had a secluded building called 'the Pavilion', it had a large glassed-in area where patients, male and female, socialised sitting on sofas and arm chairs. Surrounded by flowerbeds and blossom trees it was like a secret garden. The consultations were usually about the simple things that ordinary doctors do all the time but specialist doctors seem to have forgotten how. I had to advise on balancing blood sugars for diabetics, sorting out heart conditions prior to surgery, normalising raised blood pressure, and so on. Sometimes the surgical doctors were just trying to get us to take over some hopeless case that had somehow been admitted to their ward.

The orthopaedic unit was in its own separate area, full of patients lying in bed with their plastered limbs elevated on racks or by cranes, ambulant ones had metal pins piercing their limbs and supported by Meccano frames. Armies of physiotherapists taught patients to use their hands or how to walk again. It was reminiscent of old pictures of hospital scenes at Scutari. I went there one morning to see a patient who had ten days earlier driven his fork-lift truck into a dry-dock down by the Mersey after 'having drink taken'. He had suffered multiple severe soft tissue and bone injuries. He now was showing all the early symptoms of tetanus including the notorious 'sardonic smile' as the muscles of the face went into tetanic spasm. Lock-Jaw was a disease I recognised but had not encountered until then. After starting with the appropriate anti-tetanus serums and antibiotics I transferred him to the intensive-care-unit where he could be paralysed and ventilated. Like most working men in our city, he had a bad chest and things didn't bode well for him at all. Despite our best efforts he did eventually die from pneumonia.

Feeling I could learn no more from the books, I took the plunge and went down to London to sit the first part of the MRCP exam. This was a multiple-choice test; for each question you had to choose one out of four or five possible answers. The test ranged over symptoms and signs, lab tests, X-rays and ECGs, choice of treatments and so on. Pathology reared its ugly head again. The pass-mark was 50% but with points subtracted for wrong answers. Following the advice of the great Dr Arnold[26], I'd trained myself on specimen papers to only tick those answers I was certain of and not one more. If I could tick half or more of the questions I would have passed; if not then I will have failed unless I could find a few more answers that I absolutely knew were correct on a second run through. The examiners took a week or so to agree that indeed I had passed; so the technique had worked.

I was quite pleased now to be the only registrar at Walton with first part under their belt, others had either failed once or twice or had yet to try. Summer was coming so I rewarded myself and relaxed a bit and wishing to go sailing a bit more often I joined the sailing club at Pennington Flash.

Galadriel racing on Combs reservoir in the Pennines.

---

[26] Dr Arnold, in return for a not insignificant fee, provided advice and specimen papers to MRCP candidates. Money well spent.

Nowadays the club is set amongst landscaped grassy hills; in 1971 these hills were horrible slag heaps standing near abandoned mine shafts with their derelict winding wheels silhouetted against the grey sky. At one end of the lake was a huge ground-fill rubbish- dump populated by seabirds and nearby was a stinking shed where bluebottle larvae were grown on trays of rotting meat to be sold as bait to anglers. In Lancashire these maggots are called 'gentles'. The local fishermen would use them as weapons in the constant skirmishing between the users of the lake; the sailors on the water and the fishermen on the bank. If he felt a sailboat had encroached on his fishing ground an angler would often throw a handful of gentles high up on the boat's rig. Woe betides the skipper if his mouth was open when he looked up at the rattle as the wriggling maggots bounced of his mainsail.

I knew that the second part of the MRCP was going to be very different and also more unpredictable because it would test not only my factual knowledge but also the practical use of clinical method. In theory, if you know your job you should pass. It was expensive, about £100 in those days, to enter so failure did your bank balance no favours. I applied and was asked to come up to London in June. The bad news was that the venue was to be a peripheral hospital, not one of the big teaching hospitals. I had been warned that the consultants from such hospitals, who would be among the examiners, could be unfair as a result of being keen to appear tough on candidates, not to mention liking to show off a little. That at least was my excuse, I failed!

I do know why; it was because I argued with an examiner who was himself muddled about a tropical disease called leishmaniasis; not a condition one would ever see in England. He had corrected me on some detail and stupidly I pointed out he was incorrect himself and was thinking of babesiosis, a not dissimilar disease. Anyway, although I still maintain that I was in the right, it did me no good. A waste of time and £100.

When I returned from my trip to London and still hoping that I might have passed I found that the mysterious Mr Jenkins had relapsed into coma again. Jenkins had arrived in the ward semi-comatose, sweating and trembling a few weeks earlier. Nobody knew who he was as he had no identification with him. He was clean shaven and tidily dressed but we had no idea where he had come from, nor had the police when we had asked them to investigate. He had a tremor, his limb joints were stiff and when passively flexed they had 'lead-pipe rigidity', an appropriate simile. His face and scalp were flushed, warm and moist.

These are all signs often seen in someone taking one of the phenothiazine class of drugs which are used to treat schizophrenia. Over a week or two he gradually recovered and was able to identify himself. Indeed, he was a patient at the psychiatric unit right there in Walton Hospital. When asked, the psychiatrists refused to release his notes or any information because of some over-riding legal reason, but a junior member of their staff let slip he was a schizophrenic undergoing a trial of injections of the long-acting drug Fluphenazine, a phenothiazine.

The social workers found him a room in a nearby hostel for psychiatric patients for when he was eventually ready to leave us. As I walked into the ward on my first day back from London, there was Mr Jenkins, stiff, shaking, immobile, sweating and in coma again. Nobody knew what had happened. Gradually he made a recovery and three or four weeks later the hostel was ready to take him again, when a nurse in a beige uniform (all our staff were in blue or green) walked into the ward carrying a kidney dish. I stopped her, a syringe of Fluphenazine lay in the dish. A moment later and she would have felled Jenkins for another month. He had been prescribed the injection by the shrinks and their special nurse was going to do her duty and give it. I didn't let her near him and Jenkins escaped their clutches. My friend Charles' father was a consultant psychiatrist and I asked him to take over the care of Mr Jenkins who then lived on precariously, as schizophrenics do, as an out-patient under his care. The misuse of psychiatry described in Ken Kesey's recent book 'One Flew Over the Cuckoo's Nest' came to mind.

Just after my London exam debacle there was a jamboree at the Royal College of Physicians in Regents Park and almost all the consultant physicians in Liverpool caught the London express to attend and add their support to a Liverpool colleague, Cyril Clarke, who was to become the next College president. As promotion to consultant status was a slow process one of the senior registrars suggested that if we could sabotage the train, our career prospects might greatly improve. I privately felt that if we lost some of these old men the standard of medicine locally would also improve overnight; nothing happened and they were all back at work on Monday.

While most of us would run a mile rather that spend a night in hospital there are those in society that take the opposite view. Tommy was one of these; about twice a month he managed to be admitted to our wards. Feeling in need of a hot bath or a square meal, or perhaps just a nice clean bed as a change from his scruffy damp tenement he would go out and lie down in the street apparently unconscious. A concerned citizen would then call an ambulance and Tommy would find himself, still comatose, in Walton's AED. All the staff knew him and they all knew he was faking it, but we had to do everything possible to make sure he really was okay, which meant admitting him to the ward where after some expensive tests he would gradually recover in time for lunch. Mistaking our professional caution for timidity he probably despised us. Nevertheless, every time he appeared we were committed to going through the whole expensive routine and this was just so Tommy could get his plate of Irish-stew and a night in bed.

Tommy was lying on the examination couch beside his thick file of notes. The latest results from the lab and X-ray were there, all normal. His blood pressure and ECG were fine, his pupils reacted to light, the tendon reflexes were brisk. He lay there inert with his eyes tightly closed. This was not the first time I'd admitted him and I felt he was making a fool of us all; so I decided to call his bluff. I would do him no harm but I needed to show him that he was rumbled.

Loudly I said to the nurse that this patient really was a mystery and that perhaps the new brain-doctor from Germany, Herr Doktor Otto Von-something, might be able to help. Ten minutes later, having armed myself with a long curved intra-cardiac needle of impressive dimensions, I returned and assuming a fake German accent addressed the nurse. 'Ah seester, zis ees zee man I am demanded to see, ja? Zehr gut, now he 'az 'ad all zese tests, unt zere is only one more that we can be doing, nein? I vill now perform vot iss called un brain-biopsy with zis needle' At which point

I swished the needle's trochar insert to and fro. 'Please to be sterilising zee skin seester then I vill insert zis spezial needle through zee eear-hole and into zee brain.'

Sister swabbed disinfectant round his ear which I then held gently. Tommy lay still, but I could feel him tensing himself for whatever might come next. I then just touched his ear-lobe with the tip of the huge needle. With a cry Tommy leapt from the bed and out of the open window and was away across the lawn and off to safety. I'm not sure I can remember exactly how I wrote this up in the notes.

A few hours later Broadgreen Hospital rang asking if we could readmit one of our old patients who was in their AED and in coma. It was Tommy, I politely declined.

My happy year at Walton ended and I moved to St Helens, to Whiston Hospital where I arrived to find that once again, I was to work with absent seniors. Dr Janson had taken sick-leave for a year after a heart attack and Dr Thomas-Ellis was too busy with his private patients to bother coming to the wards very much. He was a very intelligent and kind doctor but he was a fool. He was quite unable, for example, to just prescribe a dose of digoxin for a patient in heart failure without pompously saying 'I fear that we must exhibit the leaf'. This nonsense was purportedly what an eighteenth-century doctor might have said when faced with such a patient to whom he intended to administer ground-up foxglove leaves containing digoxin. And he said it every time.

Another of his foibles was when he rang me to admit someone to the ward whom he'd seen on a home-visit or in his rooms, instead of telling me what was wrong with the patient, he would drone on about their social and family connections which were clearly of great importance to him.

I'd been there about a week running the wards by myself with only the help of two excellent housemen and with no sign at all of a consultant to brief me, when who should bounce into the ward but BB? Dr E T Baker-Bates himself. He was to be the locum-consultant until Dr Janson's return. BB's view was that Dr Janson was far too nice a man to be working in the NHS where everyone took advantage of him. He could not say no. Staff and patients took advantage of his acquiescence, the managers too. His clinics were overpopulated, there were always extra beds crowding his wards, he was overworked and he was inefficient in that he wasted time and effort on matters that others should deal with. No wonder he had suffered a heart attack. That was BB's view and I didn't argue.

BB was from a different mould and set about the staff and patients with great energy. Almoners were summoned to clear out all the long-term patients who had been blocking beds for weeks on end, prescriptions were savagely pruned, the immobile were prescribed punishing courses of rehabilitation. Our sleepy backwater of a ward buzzed as the new broom swept all before him. Next was the Out Patients Clinic; an edict was issued that all patients were to be sent back to the care of their GPs unless they clearly needed specialist care. All the stable diabetics who liked to come to see us four times a year were axed. They had resisted advice, they put on weight and would not diet, they wanted pills but probably never took them, they liked a cup of tea at the WRVS counter and a chat. BB put an end to the lot of them. Complaints were like water off a duck's back, those 'parasites' had ruined Dr Janson's health and he was putting a stop to it. I watched and marvelled, one man, well into his seventies, against the great NHS inertia and bureaucracy.

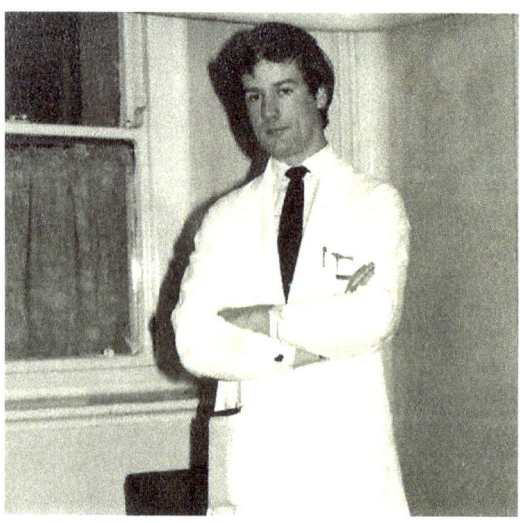

Medical Registrar. No scruffy 'scrubs' in those days.

There were great changes happening in the care of maturity onset diabetes (MOD or type 2). Despite the introduction of Sulphonylurea drugs like Daonil the death rate in MOD had not improved and a recent research study[27] showed that in fact these drugs almost doubled the risk of heart attacks and death. The greatest effort had to be made to persuade patients to lose weight, but this class of drugs actually increased appetite and made dieting even harder. The answer was to stop routinely using Sulphonylureas. The maker of one of these drugs realising the tide was turning against them started an advertising campaign aimed at us doctors and I was asked to give a talk at a dinner they were sponsoring. I was to lecture the well-fed and watered doctors on managing patients with MOD. As I had recently read the study and was impressed at how harmful Sulphonylureas were, my talk revolved round this. I don't think our hosts that night were too pleased with me and the sales manager's speech of thanks was lukewarm to say the least.

After a few weeks, BB lapsed into a manic phase[28] and started arriving at the hospital late at night to do ward rounds. He always brought some hot soup in a thermos flask that he had made himself at home and which we all had to eat at 3am. Night-sister had a firm word with him and insisted he stop doing this as everyone needed their sleep, especially the patients. Admonished and feeling put in his place BB was rather quiet as I walked him to his brand-new Jaguar 420. A bit crestfallen, he cheered himself up by proposing that we two go and visit one of his private patients nearby. I pointed out that it was still too early but he swept this aside and we drove out into the countryside to a large farm. As we drove into the farmyard, greeted by barking dogs, the lights all came on and the farmer's wife emerged looking cross. However, recognising BB, she welcomed us in for a cup of coffee. BB examined her husband and asked me to feel his spleen (he had some form of chronic leukaemia), then we left them to go back to bed. That was BB and we all had to get used to him, especially his loyal patients!

---

[27] The University Group Diabetes Program. A multinational study of many thousands of diabetic patients.

[28] BB had a mild form of biphasic disorder.

Whiston Hospital's intensive-care-unit was headed by Dr Sherwood-Jones, one of the great medical innovators of the time. I spent as much time as I could learning from this complex man who was an excellent teacher. He was interested in how to provide adequate nutrition to patients after severe trauma or infection and was experimenting with various intravenous solutions. The problem was that to provide sufficient calories to aid recovery required an inordinate volume to be infused because concentrated sugar solutions had high osmolarity that injured the patient's veins. A sugar polymer called Caloreen interested him as it had low osmolarity for a high calorific value; he wanted volunteers to test this Caloreen and I was among the guinea-pigs. After the infusion I had to collect all my urine for the next 24 hours, it was my night off and I'd been invited out to dinner and so had to take my collecting bottle with me, which rather surprised my host. The experiment was a failure since the Caloreen polymer was not metabolised by the liver as we'd hoped, and almost 100% of the dose was excreted unchanged in the urine.

This research was apparently done with no official control or sanction, nobody knew if it was actually safe. Before the era of ethical committees deciding on research protocols for safety, for consent by the volunteers and so on, a doctor could just think up something he would like to test and then get on with it. Our senior registrar, John, was writing a PhD thesis on asthma. He was interested in a subgroup of asthmatics who apparently lacked an allergic basis for their wheezing but nevertheless reacted adversely to aspirin. Needing subjects for his experiments John stalked the wards at Whiston and several other hospitals looking for asthmatics; he found one such on my ward, a man who had recovered from a severe bout of wheezing and was about to go home. Without bothering to tell anyone, John slipped into the ward and gave the man an aspirin tablet to swallow and then left to have his lunch. Our man promptly collapsed; we resuscitated him and re-admitted him into Sherwood-Jones' ICU in quite a bad way. After his lunch John reappeared and found the empty bed, he was not at all contrite and gave us all a ticking off for spoiling his experiments. Nowadays any research has to be approved by a vetting committee and this sort of wild-west approach is thankfully no longer possible.

While at Whiston I took the MRCP final exam again, successfully this time. The year passed and I moved hospitals again and became medical registrar to Dr Collins and to yet another absent consultant, Dr Watkin.

Dr Watkin was away in Europe for several weeks organising a series of lectures and seminars on diabetes for physicians in France and the Low Countries, a 'jolly' financed by a drug company. I had yet to meet Dr Watkin and without any guidance on his policies just had to improvise. During his absence a number of patients had developed problems that needed resolving. One was a quietly spoken Welshman, Mr Davis, who was in considerable pain from secondary cancer deposits in the bones of his spine and shoulders, he was dying from lung cancer. Various pain killers had been tried out but Mr Davis hated being heavily sedated by the big doses of morphine needed to supress the pain so I asked Dr Lipton, a consultant anaesthetist who ran the pain-control unit to see him. Posterior Cordotomy, an operation on the spinal cord severing some of the pain conducting fibres was suggested and Dr Lipton proposed doing this the following week. At this point Dr Watkin re-appeared looking tanned from his arduous trip to the Continent; he did not approve. He did not approve of Dr Lipton, he did not approve of cordotomy and he did not approve of registrars (me) who 'went behind his back' and Mr Davis had just to put up with his present situation, and so on at the top of his voice while standing right next to Mr Davis' bed. Poor Davis was devastated and refused to speak to me for letting him down so badly. He then literally turned his face to the wall and was dead within a week or so.

Dr Collins talked me out of resigning on the spot but the die was cast and within a few months I had left Liverpool on my way to Leeds and back into paediatrics.

CHAPTER 12. THE LEAVING OF LIVERPOOL.

On the last day of December 1973, I left Liverpool and drove through falling snow across the Pennines to Leeds, Joan waved me off. She would follow when I was settled in at the new hospital. In those days when you changed jobs you left one hospital at 5pm and appeared at the new one at 9am the following day.

I was only destined to stay in Leeds for six months. Britain had been plunged into economic chaos by the Oil Crisis and by industrial unrest on a grand scale that the politicians seemed unable to control. But despite cutbacks in the health budgets, we still had to make the NHS work.

My new boss was an Ulsterman, Ian Forsythe, his patients were scattered in wards in several hospitals around the city, and I spent my time travelling between them. The main unit was in the General Infirmary where Dr Forsyth concentrated on child-neurology; at Seacroft there was more of a mix of patients and I also looked after neonates at St James's where I had my accommodation. This was good varied experience for me and as I had been out of paediatrics for a couple of years, I had to re-learn things quickly. Paediatric neurology was a bit depressing as most of the patients were permanently damaged kids with cerebral palsy from birth, others had been destroyed by meningitis or brain tumours, some had progressive syndromes where their nervous systems were gradually shutting down. Seldom did we see a patient who was going to get much better. Many would die.

We had anti-vaxxers in those days too. Their leader was a Scottish Professor Stewart who with the tiny amount of erroneous evidence that seems to satisfy these people, had decided that Whooping Cough vaccine caused brain damage. The newspapers picked up on this and caused terror among parents who started to refuse the routine DPT[29] shots for their babies. This resulted in two effects; firstly, during that winter there were many more babies with Whooping Cough needing admission to hospital, some acquiring permanent lung damage or even dying. Secondly, just

---

[29] DPT. Diphtheria Pertussis Tetanus vaccine given in three doses to young babies.

about any brain-damaged child was immediately suspected of having vaccine damage for which the government had authorised a cash handout of several thousand pounds. Family doctors were besieged and they passed on the problem to the paediatric neurologists. Our clinics were busy with youngsters whose brain-damage had nothing to do with the vaccine but it took long hours to prove this, all of which was such a waste of time.

It took several years to stop Professor Stewart and his mis-information campaign, but eventually he was exposed as a charlatan who charged large fees to appear in court, especially in America, as an expert witness in medical negligence cases. That sorted out, everyone went back to having their DPT shots.

One evening I was checking on my babies in the neonatal ward when the houseman from the children's unit on a lower floor rushed in. The duty paediatrician was at Seacroft and unreachable, would I help as they had a real problem? A three-year old boy had been admitted with extensive bruising and the doctor in casualty had called the social workers believing that this was a battered child. The paediatric houseman realised that the bruising was in fact due to septicaemia and not trauma and that the boy was very seriously ill with a high fever. The social workers, bent on child-protection, refused to let the houseman near the child and he was beside himself with anxiety. I had no responsibilities in that particular ward but we hurried together down to the unit to find a policeman questioning the parents while the social workers were about to leave, apparently intent on taking the child with them.

I'm not sure what I said to them but I reclaimed the boy and started treatment. Meningococcal septicaemia is often fatal and it was touch and go for several hours. I tried to rescue the parents as well but they had been taken off to the police station by this time. Eventually I found the consultant in charge of the ward but even he could not sort out the parents' plight and they weren't released for another 24 hours. They must have been sick with worry over their son, and completely powerless once official wheels had started turning.

Violence against children is a very serious matter and is often hard to detect, all health workers should always be on the lookout for signs of it;

but there are more causes of bruising than trauma, including leukaemia and infections. Thorough assessment of the child including admission to hospital needs to come first before accusing parents of harming their child. This little boy would probably have died had we not insisted he stay in the ward.

This enthusiasm on behalf of some doctors and social workers to disbelieve and accuse parents may have been due to the influence of Dr Roy Meadow who was a consultant paediatrician in Leeds. Dr Meadow was very interested in child abuse and was famous for introducing the world to the Munchausen Syndrome by Proxy[30] a few years after I knew him. He was also the scourge of parents whose children had suffered cot death. The so-called Meadow's Law said that two cot-deaths in a family is suspicious and three is murder. He was much later accused of giving inaccurate evidence in court and disgraced, although his name was cleared by the Appeals Court in the end.

A few weeks later Joan and Diane arrived in Leeds and we all moved into one of the hospital's married quarters. It would be a short stay as by then I had started to think of a relocating to Hong Kong where I had been offered a job. After ten years as student and doctor in the NHS I was going to leave, not without many regrets. The NHS had given me the chance to become the sort of doctor I wanted to be; its doctors and nurses and its myriad patients had allowed me to learn my profession and now I was leaving for a new place. For me the NHS represented the best way of delivering the best care to the largest number of people in the most efficient way, but I was not unaware of its faults. It was underfunded and it was increasingly bureaucratic; and this was before all the even more restricting Health and Safety regulations which were introduced a few years later.

Joan, Diane and I flew to Hong Kong at the end of June to start a new life in the Far East and that is another story.

---

[30] Munchausen Syndrome by Proxy. Where a parent poisons or injures their child to fake an illness and thereby gain some attention for themselves. Many of us doubt the existence of this as a separate phenomenon and see it as hardly different from other ways of abusing a child.

BOOK 3. HONG KONG.

CHAPTER 13. DOCTOR ORAM & PARTNERS.

Seeing patients is very similar anywhere in the world, even though no two patients nor their doctors are the same. 'What seems to be the trouble? How can I help you? When did this first start?' Then you're away: clinical-method: history, examination, differential diagnosis and some tests perhaps. If you don't have a good idea of what is the matter within the first few minutes, you probably never will. Then: advice, treatment, prescription. Often the laying-on of hands and some quiet advice is all that's needed as most illnesses heal themselves without interference. There are no side effects from that. Other times the doctor needs to act urgently. Accurate diagnosis and a good nose for problems are of the essence, as is the ability to sum up your patient as a person. After all, you are treating a person with a disease, not just a disease alone; both person and disease must be well understood.

So, in they came, every ten or fifteen minutes. I was the new doctor and on display as much as they were. As I assessed them, they assessed me; I knew that these were important encounters that might determine many years of relationship between patient and physician. Some patients would become firm friends while others felt no need to invest much in our exchanges.

Trained as a paediatrician, I was in the habit of addressing the patient, the child, while politely acknowledging the accompanying parent. With babies it was more veterinary but even with toddlers I felt the need to make them feel they were the focus in the room and to encourage them to speak and act for themselves. I did not like to baby them and I absolutely refused to give them sweets or stickers as a reward for their co-operation. We had toys in the waiting room but otherwise the clinic was business-like.

Children are not stupid; they know doctors sometimes have to give injections or make them undress and they are not going to be reassured by candies or Disney characters on the wall-paper. My approach was to be honest, to assure them that if it hurt it would be brief and no more than they could stand. I never told them to be brave, preferring to just ask them to be 'tough' for a moment. Of course, there were tears but I tried to avoid resentment from, 'You said it wouldn't hurt, but it did'. Instead, I concentrated on good technique, quick firm action, getting it right first time and complete honesty. Nurses were forbidden to say that 'it won't hurt.' Children can spot such dishonesty and soon wonder what else we might be lying about.

At the end of the morning clinic on that first day, the electricity went off. The whole district had a power-cut. Before I could leave for lunch, a parent rang demanding a home visit to a nearby flat in Bay Court claiming their child was too feverish to be moved. I drove over to discover there was a 12 storey climb to their flat as of course the lift wasn't working during the power-cut. The stairs were steep, dusty and hot. I arrived considerably out of breath and almost dripping with sweat. The patient was cheerfully running wild round the apartment. The ear-ache was dealt with quickly and then I gave Mum a piece of my mind for demanding what seemed an unnecessary visit just so she could avoid having to use the staircase herself. Doctors need to stand up for themselves even on their first day at work or they will soon forfeit their patients' respect and they will think you are prepared to be at their beck and call.

This was to be my life for the foreseeable future. Clinics, visits to patients in hospital or at home, almost always on call. I was to have 'privileges' at three private hospitals, meaning I could admit patients to the wards and care for them myself or ask specialists for their opinions and help. I could send my patients to the lab, the pharmacy or the X-ray department as well as use consulting rooms in the out-patients area. My new partner, Bill, as a surgeon had the right to perform operations in these hospitals but I was to find that not all the doctors with operating rights were actually fully trained and there was very wide range of abilities amongst my new colleagues.

In England's NHS, the career ladder is clearly defined. Doctors move up the scale as they gain experience under constant assessment from their seniors. Professional exams have to be passed and some research published by the doctor or surgeon in training until eventually, if he comes up to the required standards, he achieves 'consultant' status and begins to work unsupervised. In Hong Kong this was also the system in the government's hospital service; however, specialists not quite making the grade or perhaps passed over because they didn't quite fit in, could set themselves up in private practice where there was little oversight or supervision. Some were superb, others less good, a few actually dangerous. Some excellent doctors, including several professors from the University, on retirement from the very busy and understaffed government hospitals, moved into the private sector where they could continue to work with less pressure; not to mention make good money to save for their twilight years. With these specialists whom we could call on for their opinions, we private doctors could provide our patients with the best care and know that expert help was only a phone-call away.

One old man, a famous surgeon locally with a large and devoted following was often sitting in the doctors' room next to the theatres. Generally, he lounged there, chain smoking and reading a Chinese newspaper. He was fully garbed in his 'scrubs': apron and white wellingtons with a cotton mask hanging loosely round his neck. From time to time, he would stub out his cigarette, pull up the mask and nip out to the theatre entrance area to shake the hand of a patient already lying on a trolley with an IV running. After a few encouraging words, they would disappear together into the operating-room. When the patient was asleep, the old man would return and light up again while his 'junior assistant' would competently perform the surgery. Observing the old man, I couldn't avoid noticing how his hand shook as he held his cigarette; this tremor had ended his surgical career but he was still able to make accurate diagnoses and provide support for his patients. It is easy to be judgemental but he had found a way to continue practice and provide, if somewhat dishonestly, care for his loyal patients.

As it was summer when I started work, many expatriate families were away in their home countries and not likely to return until the school year started two months later. This meant the clinics were quieter than usual though the hot and humid weather brought its own problems. Plenty of youngsters with swimmer's-ear for example, a painful condition entirely due to soaking in chlorinated water day after day at the Country Club pool. We saw sweat rashes and sunburn, insect bites too, but not so many coming in with chest trouble or asthma but the routine of baby care with developmental checks and vaccinations kept me busy and before long I was to help at the delivery of my first Hong Kong baby.

The Canossa Hospital in the mid-levels is a charming place, with its front drive overhung by mature trees, among them the dramatic red blossoms of the Forest-Flame. From its wide windows, the Harbour was to be seen shimmering far below, busy with shipping. The sound of traffic rose from Central District with the rattle of piling hammers from building sites in the city. The hospital was run by an Italian order of nuns headed by their fearsome mother superior, Mother Jo, who turned out to be the kindest person you could ever meet. Most of the nurses were local Chinese women although among the senior sisters most were nuns with their wimples and headdresses. Apart from a crucifix above every bed in the wards, religion otherwise seemed remarkably low-key. Most doctors are atheists in my experience, though when believers, they do tend to be devout, even fanatical.

Bill preferred to base himself at the Canossa as it was the nearest hospital to his home and also it had a good car-park where he could leave his Jaguar during the day. The maternity wards were on the second floor with the operating theatres on a floor above. While I was not usually present for uncomplicated deliveries, I had to be there when problems occurred, especially for Caesarean Sections, so I spent a good deal of time waiting around on those two floors. The patients were not as enamoured with the Canossa as was Bill and found the brisk and starchy nuns in their elaborate head-dresses less approachable than the English ladies who nursed at the Matilda where the food was reputed to be better and the mountain breezes brought a certain cool in the summer. The Matilda is high up on the Peak and has a colonial style main building with deep shaded balconies overlooking the China Sea and its islands. Myself, I would not care to judge the comparative merits of the two hospitals' versions of that style of cuisine called 'Chinglish Cookery' which combines the worst features of European and Asian methods.

The Matilda's maternity unit at the time was actually in a separate building beside the main gates and it lacked the sea and mountain views of the main hospital, so the young mothers were often disappointed with their stay in the gloomy private rooms overlooking the maintenance area.

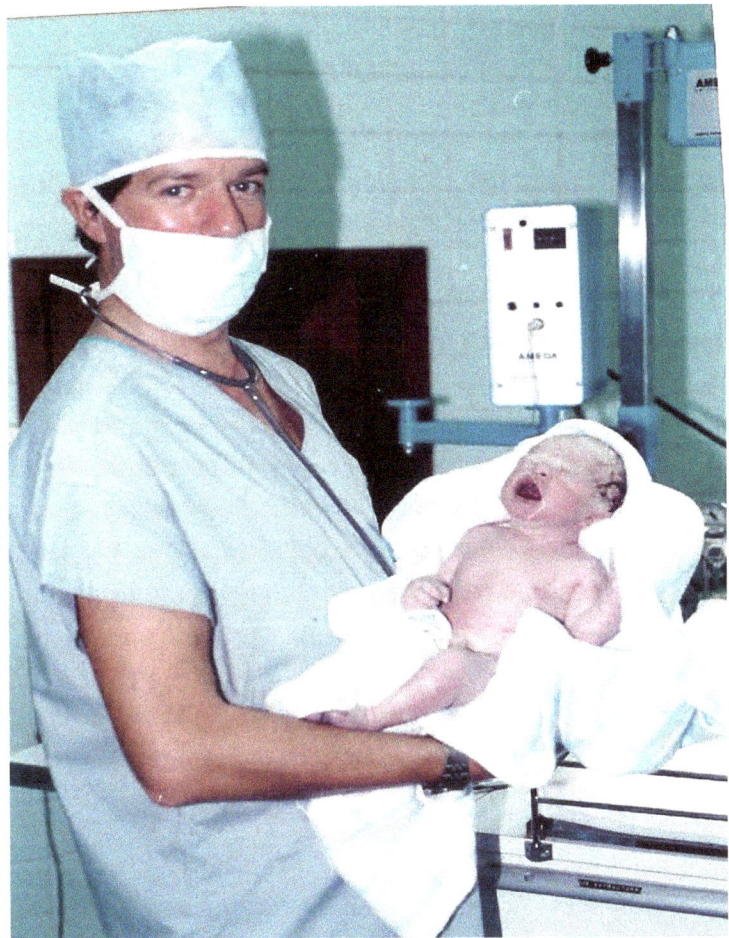

Matilda Hospital delivery room, this new arrival looks healthy enough.

One big difference between the two hospitals, was in their attitude to the fraught issue of feeding the babies. The Matilda's maternity unit had been taken over by a coven of midwives convinced by the Breast-is-Best movement and woe betide any young mum who even considered the advantages of bottle feeding. Such heretics were hectored with almost certainly apocryphal 'scientific' reasons involving immunity, bonding and whatever. Colostrum was the elixir of life, while formula was the milk of the devil. I never managed to persuade my nursing colleagues that lactation in the human took a few days to get started, unlike lower animals which secreted milk even before starting in whelp. Drowsy babies were taken to the breast at all hours, no one had any rest, nipples cracked and most mothers were in tears. The only hope for these women was an early escape home where the milk promptly 'came in' and everyone calmed down and could think straight.

At the Canossa they had other ideas. The nursery was run by a gang of redoubtable Chinese ladies of a mature age all dressed in their uniform of loose white chemises above black silk trousers and referred to by the honourable term of 'amah'. The amahs knew exactly how babies should be fed. Every three hours the infant was lifted from his bed, unwrapped and diapers changed. A bottle of milk was then inserted in the correct orifice and was empty within ten minutes. The child was then swaddled tightly and laid on his back neatly positioned in a straight line with his fellow neonates. No baby existed who could resist the determination of these good women. Mothers were not to be bothered as they needed to 'rest'. Some of the better-off local tai-tais found this all so convenient that they chose to leave their new babies under the care of the amahs for several weeks after they had left the hospital themselves. A thoroughly modern mother had quite a battle on her hands if she wanted to breast feed, especially if she wished the baby to be in the room with her and comfortably sprawled on his tummy in his cot.

This is how I found the situation in 1974 and I never chose to interfere although I do admit to preferring natural breast feeding for my patients as it does seem more natural for a woman to put her bosom to its intended purpose when given the opportunity. I do not know how today's nurses, paediatricians and parents handle the vexed question of infant nutrition but I am quite sure that their most firmly held views still won't really bear too much scientific scrutiny.

CHAPTER 14. GETTING TO KNOW HONG KONG.

When we arrived in the Colony, I really had little knowledge about China and Hong Kong. About the Opium Wars I knew nothing, though no doubt they had been mentioned in school lessons but they were long forgotten by me. In Leeds, hoping for enlightenment about my future home, I had visited the local library and found a book by Somerset Maugham which proved useless though enjoyably dated and also Jean Gittins' *Eastern Windows, Western Skies,* her memoir of the early 20th Century. It had only been published 5 years earlier and began to give me some idea of Hong Kong before and after the war, including the horrors of the Japanese occupation.

Bill Oram had mailed to me in Leeds a copy of the Hong Kong Government's latest annual report containing beautiful colour pictures of Hong Kong. Photos of young women water-skiing and riding bikes made it look leisured and very attractive, as did even the photos of buses and new railway engines, not to mention dramatic views of the construction of a huge new reservoir being reclaimed from the sea. The text however was a dreary iteration of expenditure projections and reports of government committees and departments.

A few weeks after our arrival, the excitement and novelty of the move to a new place began to fade, I suppose we were a bit home-sick and realising that although we had left many problems back in England, there were plenty of new ones to be solved here in Hong Kong. Despite invitations and outings with Bill's social circle, we felt a little lonely as they were all a bit older than us. Our leave-flat was rather cut off with no public transport and many of the neighbouring flats were empty, their families escaping the summer heat by taking long leaves in their home countries. This was harder on Joan than me and I knew she missed the camaraderie of the workplace. This of course proved to be a temporary phase and it was not long before our social life was booming as she met other mums at the school gate and I received invitations from patients to their homes for dinner or to go on picnics. Some of them were to become lifelong pals. Later when we joined the Yacht Club, we found many new friends there.

Joan and Diane on the bund getting to know Aberdeen

Nevertheless, I lacked a feeling of connection with Hong Kong as a place where major changes and events had happened and which had a place in history. Coming from England, where almost every street corner has its story, I found Hong Kong somehow meaningless and I lacked the necessary foundation in its time and history. I expressed this to a schoolteacher acquaintance who completely understood and suggested that I should read the book *Foreign Mud* or, if I liked a racier read, James Clavell's *Taipan,* and learn about the Opium Wars and the founding of the Colony. I devoured both books and began to realise that my new home was in fact full of history, glimpses of which I began to recognise as I explored the town. Street names became significant and company names were linked to traders from the buccaneering days of the early eighteen hundreds. I followed up by reading Harrison Salisbury and Jonathan Spence's works about Chinese history and started to really feel settled in this new place that had so recently emerged from an ancient civilisation, its birth as traumatic and exciting as any new nation. Hong Kong was beginning to be home and as time passed, I became hungry to know more about Chinese culture and history and even to learn the language.

Although we could not yet afford to buy them, we came to love looking at and handling the antiques in the shops along Hollywood Road and Cat Street. Several friends had collections of porcelain and an orthopaedic surgeon colleague tried to encourage me to appreciate old snuff-bottles. Tony da Roza, a doyen of the local Portuguese community, was a great worrier and was convinced that the Chinese were poised to cross the border and drive us all into the sea. His wealth seemed to have been concentrated in two carrying cases in whose padded depths lay his entire collection of the exquisite little bottles, ready for flight to safety when the PLA rained Armageddon on us all.

Some years later, in the 1980s, when we felt we understood something about Chinese art we joined a group[31] supporting a small museum at the Chinese U. Joan eventually was voted on to a committee that raised money to buy ancient artefacts to display there. Peter Wu, the museum's curator, used to take groups of members round the antique shops and into their back rooms. Peter would explain the valuation of items, how to recognise the special glazes and spot fakes and copies. Later as China opened up to visitors, we joined the group's tours on the mainland and gained more still of an insight into China's art, its history and culture.

Like me, many of my patients were expatriates living far from family and close friends who would, in their home countries, have rallied round when a new baby was born or someone fell ill. Many new mothers deeply missed the support their own mother would have provided 'back home'. The concept of 'paternity leave' was quite unknown and employers were not keen to give the new dad much time off work. Of course, many grandmothers did come out to Hong Kong to welcome their new grandchildren and help, but all too often the new parents were by themselves and leant heavily on their doctor for support.

---

[31] Friends of the Art Gallery, later renamed Friends of the Art Museum, Chinese U.

Their lack of experience and support led parents to worry unnecessarily about their babies' behaviour which their own parents would have taken in their stride, surrounded as they had been be by an extended family. As an expatriate new mother left the comfortable environment of the maternity ward, the responsibility of caring for this new life could crush her. Post-natal depression was no less common in the 1970s but was much less recognised by families and by health professionals than it is today.

I became used to tearful mothers worrying about small problems that would certainly sort themselves out. Difficult feeders, babies who wouldn't sleep, babies with colic or who regurgitated their milk. Some parents were concerned over their baby's poo, its colour or frequency. In the days before disposable diapers, there were frequently nappy rashes that grandma would have sorted out in no time, but grandma was thousands of miles away and there was no e-mail or Skype. There were no young mothers' groups on the net. There was no net! A lot of the support had to come from me.

Coming from the NHS, I was surprised that in the private system there was no such thing as a well-baby clinic where experienced nurses could help to sort out these everyday problems; virtually non-problems. So, I arranged that one afternoon a week, when there was no doctor rostered to be there, our nurse at Repulse Bay would run such a clinic. It was to be free of charge and open to all comers. It offered advice and simple remedies only, so for anything requiring medical treatments or vaccinations they had to see one of the doctors. This was immediately popular and even Bill, who was resistant to the idea of giving away anything free, came round to agree that it was at least an effective public relations strategy.

After three months, we moved from the leave-flat near Stanley into a delightful but small top-floor flat with floor to ceiling front windows overlooking a communal garden in Shouson Hill. There were four blocks of three storeys and we were all kept organised by the caretaker, Charlie, who lived in an open space under one of the blocks. He had several garden tables and many outdoor chairs arranged within a circle made of old fridges and air conditioners that whirred away day and night keeping Charlie and his drinks cool. Who paid for the electricity I have no idea, certainly not Charlie.

Having been shocked at our first electricity bill, we cut down on air-conditioning and relied on ceiling fans to move cool breezes from the open windows. The local insect population benefitted from this and we all itched from their bites. Joan bought some mosquito coils and we hung the green spirals round the flat but they seemed to have little effect. Jenny from the flat downstairs, seeing them hanging there, laughed and told us that they actually needed to burn and smoulder to release their insecticide. It was hard to pretend we knew this already; another thing to learn. Jenny's daughter, Katie, and Diane became firm friends and the girls were up and down the stairs all day. This suited Diane as Jenny was very generous in handing out sweets and chocolates to the girls, whereas Joan was very strict; Diane was only given a treat on Friday evenings.

CHAPTER 15. DOWN TO WORK.

I walked down the steep pavement on Wyndham Street in the warm morning sun. It was so steep that the path is stepped in places. I was trying not to inhale the awful smell from a roadside stall selling deep-fried stinking-bean-curd snacks to the schoolchildren dragging their heels on their way to the nearby Catholic school. I had left my car, our nearly new Austin 1100, at the Canossa, and after reviewing several newborns on the second floor, I was on my way to the clinic down in Central. I liked to walk down to Central so I could observe Hong Kong life unfolding at the start of the day before the heat became intense, I could watch the children in their immaculate white school-uniforms larking about, old men in pyjamas taking their cage-birds for a walk, 'black and white' amahs gossiping on doorsteps, a laundry with its doors open emitted steam and the smells of ironed sheets. There was Damask House displaying bolts of fine cottons and linens through an open door peeling with faded Lunar New Year decorations. Chauffeurs leant on their Mercedes and Holden limousines having a smoke. The gweilo doctor walked to work carrying his bag; just another playing his part in the morning scene.

South Bay Road; our clinic was on the first floor above
Park'N Shop.

The waiting room was crammed and people waited outside in the corridor too. 'Why you not come early doctor? Too many people no appointment now', cried Vivienne. It took a while to clear the chaos. I had only been in the practice a few months and I was already busy. Hong Kong was recovering from both the worldwide recession and also the aftermath of the Star Ferry riots[32]. New expatriates were arriving daily from all over the world. The three stock exchanges were registering astronomical rises with equally dizzying falls to match, the business world was rushing in to take advantage, and our little practice was dragged along in the current.

No time for leisurely lunches now as I rushed between the two clinics and fitted in hospital calls too. Never say no, said Bill, or they won't bother to ask you again. The key to successful practice was the three As: availability, amiability and ability, in that order. The first A is the most important; your peers may well judge you by your clinical ability but it's the patients who count, so unless you are available, the other two As just don't come into it! To that I would add a fourth A, action. Patients expect you to sort out their problems, they want a quick resolution. When you can competently perform a procedure yourself, you should get on with it and not procrastinate by sending them to a specialist or for unnecessary tests 'just in case'. The Australian author Shirley Hazzard put it succinctly, 'Respond[ing] emotionally rather than pragmatically [is] a device to retain a sense of patronage'; it's no good to just empathise with a patient as that will only make them more dependent on you. That may be good for your ego but it does not help the patient.

More serious cases had started coming my way, the humdrum of minor injuries and infections now leavened with patients with major conditions. Sam Cohen was a big man, though not as big as his wife. A successful Jewish businessman, prominent in the rag trade, he had a big flat and a big car and a big personality. He had survived a coronary thrombosis some years before. He rang me asking for a home visit one breakfast-time as he was short of breath and feeling very ill. As I arrived, he looked grey and his pulse rate was down to around 30 per minute; it was clear that he was in heart-block. I needed to get an ECG, he needed a pacemaker urgently, nothing

---

[32] Star Ferry Riots of 1966-7 at the time of the Red Guards in China.

else would do. The ambulance arrived, Sam was given oxygen and his ECG showed complete heart-block as I had expected. Sam was in trouble, so I jumped into the ambulance with him and, with bells ringing and blue light flashing, we nosed into the morning rush-hour traffic only to come to a sudden halt slewed right across Magazine Gap Road. With the ambulance's steering broken, we were stuck. The traffic, also unable to move because we had blocked the main road into town, built up and the replacement ambulance could not get through the jam. There was nothing for it but to move Sam into my little car. By driving over the kerb, we just squeezed past the dead ambulance and I drove Sam to the hospital myself. Without oxygen, he was looking greyer by the minute until we eventually arrived at the Canossa where I could insert a temporary pacing catheter and get his heart rate above 70. Three weeks later he was back at work, loudly proud of the pacemaker under his left armpit which everyone was expected to admire.

I started to see a few patients on the Peak on two mornings a week at the Matilda Hospital before heading to our own clinics. Davy, a boy of about eight turned up one morning with a discharge from one nostril. His father said this often happened and generally settled with a course of antibiotics, would I oblige as his regular doctor was unavailable? Davy was not pleased when I asked to examine him and struggled wildly at being touched. Nevertheless, I managed to see inside his nose and could see something shiny deep in the nostril from which smelly green mucus was leaking. Further investigation was needed and a few days later he was anaesthetised and Bill removed a shining red and gold coat-button from his nostril. His mother recognised it as one which had been lost when Davy was a toddler.

Foreign bodies in the various orifices of youngsters are a very common problem and I was often removing bits of toys, nuts or raisins from ears and noses. One little girl's alarming emerald-green vaginal discharge proved to be from a fruit-gum she was keeping in a safe place.

Jason was three and had just arrived from England with his family and had developed a cough during the flight. He was seen the evening before by an emergency doctor who had taken an X-ray and later phoned the family to say no treatment was needed. But Jason's cough was in fact getting worse; what did I advise? I rang the radiologist but could only speak to his secretary, who said that he had taken the X-ray plates away with him intending to show them at a lunch-time medical meeting as a case of a rare condition the secretary thought he had called 'Swyer-James Syndrome'. I explained to her that the child's condition was getting worse and I must see the films urgently. The radiologist eventually called me back but by then events had resolved the matter

I had never heard of Swyer-James Syndrome and went to look it up to learn that it was an incidental finding in which one side of a chest X-ray looked darker than the other, it was usually evidence of previous lung trouble but was otherwise of little significance. To my mind a much more likely cause of such an X-ray appearance however would be an inhaled object like a peanut obstructing the airflow in a bronchus. Jason came over to my office and I explained all this to his father and proposed that we hold the boy upside down and give him a few thumps on the back of his chest in an attempt to make him cough and dislodge the peanut or whatever it was. In the event my diagnosis was wrong; it was not a peanut but a pistachio that Jason coughed up almost immediately on being hung head down. His elder brother then told us that on the plane as their parents slept, they were each given a little packet of nuts by the flight attendant. I debated whether I should ring the radiologist and warn him against giving his presentation; Swyer-James Syndrome had suddenly become even more uncommon.

Bill was involved with SARDA, the government's drug rehabilitation service and was also managing a few addicts in our clinics; inevitably some of them spilled over into my sessions despite my firm intention to avoid them. I found them most frustrating to manage. Some had to be admitted to hospital to dry out, to 'de-tox'. Even there the dealers still found them. They would creep up the back-stairs and slip into the private wards. Parents were paying out thousands of dollars for their darling children's health, while the kids were secretly getting their fixes right under our noses. They traded urine samples, they stole money, they sneaked out at night - what could a doctor or a parent do?

It is all very well to take a moralistic standpoint when dealing with addiction but physicians are there to help the sick without any reservations just as in warfare a doctor will treat an enemy soldier; and so a physician faced with an addict who is lying and cheating should still give of his best to help, whatever he may feel privately.

I had gained some experience with helping alcoholics when I worked in Liverpool and understood that someone who has an addiction will make every effort to preserve that addiction. The drug completely takes over their life and their personality; everyone knows this but when faced with someone you love who has become addicted, it is nearly impossible to make those changes in the relationship that are needed to rescue that person. Parents cannot believe their son or daughter can be so devious and untruthful. They do not realise the addict has lost all conscience, has no moral compass and is in thrall to their new master whatever it is; heroin, cocaine, alcohol. The addict will use anyone and anything to continue, to get the next fix.

Angela was about fifteen, a pretty girl from England, spending three years in Hong Kong while her father managed the local branch of a large international bank. Until recently she had been doing well at school-work and sports. But for a few weeks she had been unwilling to take part in family activities and her grades at school had fallen. She just wanted to loll around at home. Her parents had put this down to an adolescent phase but the school now wanted to suspend her, along with some of her friends, suspecting them of drug taking. The family were sure it could all be cleared up with a letter from me. They felt confident that a urine test would clear her name and she could remain in school with a clean record. Angela agreed to cooperate and pull herself together and said she would submit to urine tests, 'if she must!'; shrugging her shoulders as she said it.

Then came the problem. Mum produced a small jar containing an amber liquid asking that we test it 'and that will be an end to it, won't it darling?' When I said that for narcotic testing, the urine must be passed into the bottle before a reliable witness and that their specimen just would not do, they protested vigorously despite my continuing insistence.

I knew that they left the clinic extremely angry with me. I felt they were 'shooting the messenger'; but I realised that their little girl had never lied or cheated before and it was impossible for them to accept the change in her that was so obvious to her teachers and me. I should have done better but it might have taken weeks for the truth to be brought home to the parents, in which time she could have come to all sorts of harm. Several youngsters I knew, all from good families, died from overdoses and others ran away from home to live rough. In my experience, the only course was to keep these youngsters pretty well prisoners for quite a long time; no outings, only a few chosen friends and home tutoring. In reality the best method was to send them away from Hong Kong with its easy availability of drugs.

Heroin was and is so easy to obtain on the streets of Hong Kong, a small packet costing hardly more than a pack of Marlboros. Older children peddled the packets even inside the schools. The 'Shack' was a makeshift café on some derelict ground behind the flower stalls on the corner of South Bay Road, right behind our clinic. It served cold sodas and bowls of noodles to many of the children from the nearby school at lunch time. It had been raided by the police a few times looking for pushers who hung around trying to sell heroin or marijuana to the kids, but it stayed open and was popular even with the children who didn't use drugs as the Chinese food was cheap and tasty.

As time went by, I became more relaxed in dealing with addiction problems despite the frustrations, but it was a sad business especially when an accidental overdose took the life of the teenage son of a good friend.

It was not only young people that became addicted to drugs but adults too. In the 60s and 70s Diazepam (Valium) was very widely prescribed for many conditions, ranging from mild anxiety to back-ache, from spastic-colon to mild depression. It was very effective. Anxiety would vanish or muscles would relax, allowing a sore back to settle down. The problem was that when you stopped taking it, your symptoms, especially anxiety, usually recurred. Then progressively larger doses were needed. Alice, a housewife and part-time secretary, asked for a prescription for Valium 10mg, the highest strength, which she had been taking for several years since suffering a minor nervous breakdown. I wrote her a prescription for 30 tablets with a warning to cut down her dose. The local chemist called me later asking whether I really wanted her to have 300 tablets which she rightly considered a risk for suicide. I asked Alice to come back on a pretext. She confessed that she had over a thousand Valium and Mogadon tablets in her handbag; enough for several suicides. I challenged her and offered a way towards stopping her drug abuse, but of course she would not cooperate and I never saw her again.

Valium is much less used today because the risk of dependence is well recognised, the current problem revolves round the use of codeine derivatives for chronic pain. Used for back-ache and other long-standing pains, it works well initially but then patients find they need ever increasing doses for less and less relief. Inevitably overdose and accidental deaths are the result. In deprived areas, even of developed countries, overdoses of these opiates are a leading cause of death.

When I had been in Hong Kong for several months I decided to drop in and meet the pharmacist at Watsons-The-Chemist in the old Central Building. We had spoken on the phone many times about prescriptions but so far had never met. She was called Julie-Anne and had a strong Australian accent. For some reason, certainly based on her voice over the phone, I envisaged her as a well-built lady with red hair. I could not have been more wrong. Julie-Anne was in fact a diminutive young Chinese woman with a jolly smile and Harry Potter glasses who had trained in Melbourne.

It was not only the teenagers who caused problems in the family. Anne was about 15 and lived in one of the old low-rise blocks of flats in the lane behind our clinic in Repulse Bay; a sensible girl, she planned a career as a nurse. She worked a few hours a week for us doing clerical chores in the clinic. She would sort out case notes, write receipts and file lab tests to earn some pocket money. One evening she was reluctant to leave for home and admitted that her life was very stressful due to her parents' heavy drinking. They could be violent; the house was unkempt and their shouted arguments disturbed her sleep and interfered with her studies. In fact, it was she who was now running their home; doing what cooking she could and cleaning up after her parents drinking bouts including their vomit, the broken furniture and the empty bottles. She was at her wits end. I let her have a key to the waiting room just to escape and to do her homework in peace. Al-Anon, a charity associated with Alcoholics Anonymous, gave her great support and mentoring. She did pass her A-levels a few years later and left Hong Kong for an English nursing college. She kept in touch after she qualified and eventually married and started a family far away from her parents.

The taxi rank where I waited for a cab back to my car, parked up at the Canossa, was outside the Pedder Building on the opposite side of the street from the Central Building. As we huddled out of the midday sun in the shade under the arches, a beggar-woman worked the queue. She was there every day looking pathetic and holding a baby in her arms. This baby, wrapped in rags, always had house-flies crawling on its dirty face. I sometimes gave the mother some small change until a tall Chinese woman beside me in the line remarked that I should not be too affected by the sight of this beggar who, she said, smeared the baby's face with syrup and had brought the flies with her in a matchbox. There were several beggars in that area including a blind man with a fixed smile selling chewing gum from a tray and another man with enormously swollen legs and feet. I could see he had tight tourniquets tied above both knees but I thought his was a good act and occasionally gave him a few coins for his efforts.

At that time, the government paid a small subsidy to out-of-work people, it was called the anti-begging allowance and was supposed to keep them off the streets. Without the support of a proper Welfare State, the people at the bottom had little choice but beggary. One sight that still makes me sad, even today, is of an old lady bent over and pushing a cart laden high with sheets of cardboard for recycling. These ladies, always ladies, are to be seen everywhere.

The third hospital that allowed me privileges was the Adventist on Stubbs Road. It is about half way between our clinics in Repulse Bay and in Central, so I would drive past it at least twice a day. It was run along American lines by the 7th Day Adventist church and was very modern and forward looking. The big downside was that the kitchens were strictly and notoriously vegetarian and my patients just didn't want to eat the meals they produced. I used to note that in the evenings, many other patients must have taken the same view as the corridors became littered at mealtimes with packaging from take-away food outlets. Apart from the catering the hospital was excellent and the nursing staff reliable and knowledgeable. The place was busy and swarming with doctors, so I knew that if anything went wrong, there was plenty of expert help for my patients if I was held up on my way to the hospital.

The senior physician, a Canadian cardiologist, was very energetic and a great believer in the benefits of exercise. He founded a Running Clinic and on Sunday mornings he had large crowds of people in the hospital car park all dressed in shorts and t-shirts, having their blood pressure and cholesterol checked before going on healthy runs round the hills behind the hospital. For me, exercise for its own sake is anathema and my own patients could relax in the certainty that I would never ask them to work up a sweat on a run or in the gym.

That is not to say I encouraged laziness, on the contrary, activity of the mind and the body was to me the essence of a healthy life. Patients with symptoms of stress were common enough and the pressures of Hong Kong life tended to narrow my patients' activities. My usual advice to those in my care was to take up an interest, something to exercise their imaginations and their bodies. Something to balance the pressure of working the long hours expected in the Hong Kong workplace.

CHAPTER 16. CARMEN.

If it seems that I led the life of a slave to my work, nothing could be further from the truth. We would not have mobile phones for at least another twenty years but somehow, we managed to be within reach for calls from hospitals or patients. I carried a bleep but Bill resolutely refused one. Patients ringing his home would often be informed by Ah Fong, his venerable black and white garbed amah, that 'Massa' was 'topside' meaning at the Matilda Hospital on the Peak, or perhaps, 'Massa, he go Shek O' meaning to the golf course where there was a telephone at every tee. Either the caller managed to interpret this or they bleeped me. Hong Kong local calls are free so I could phone back from anywhere: shops, restaurants, the cinema; nobody minded you using their phone to make a call. The one place from which I could not call was out on a boat, and in crowded hot Hong Kong, out on a boat was often the best place to relax. Joan, Diane and I were often invited on friends' boats. The owner or his Chinese 'boat-boy', usually an elderly retired seaman, would bring the junk to a nearby beach or more usually to a jetty or some slippery waterside steps where we would precariously jump across the gap clutching baskets of food, drinks and bathing costumes; children close behind.

Boat trips were usually to outlying islands with clear water and sandy beaches, where we would cool off in the sea and drink San Miguel beer in the shady cabin; the ladies would serve picnics, the men blowing up water toys for the youngsters. Bronzed excited children plunged from the top deck into the sea. The little ones could hardly swim but they struggled to doggy-paddle back to the stern platform and then ran, screaming with delight, back up the stairs to do it again while mothers urged caution, sun-cream and the wearing of hats.

On one trip our host's somewhat feral older children had rowed ashore to explore the sandy beach. The eldest boy returned carrying a human skull still with some scalp hair and leathery skin clinging to it. 'Look what I've found. It was in this stone jar by the beach'. He had clearly come upon a traditional Chinese burial shrine where the bones of much-respected family members had been put in earthenware pots and placed on a hillside overlooking a beach or a stand of trees. This is thought to bring good fortune to the relatives who have gone to this bother and expense for their forebear. The rather smelly relic had to be surreptitiously returned to its resting place without upsetting everyone on the beach.

Usually, we would raise the anchor later in the day and motor to a village in a nearby bay where fish-restaurants stood on sturdy bamboo stilts above the water and there was a rickety pontoon where we could land. The place we liked best was called with all due irony, the 'Lamma Hilton'. A dozen mismatched tables were crowded with happy diners calling to waiters to bring ice-cold beers from the rusty old fridges. After haggling over the prices for a live fish from the tanks of bubbling sea-water near the kitchens, where huge woks of rice and vegetables were cooked over roaring paraffin flames releasing tantalising aromas. The chefs banged their utensils and yelled through the steam and heat to one another in Cantonese. We would eat steamed prawns by the catty, curried crabs, squid tempura, clams in black-bean sauce and greens steamed with garlic and soy-sauce. Then the highlight of the meal arrived: the garoupa we had chosen. Now steamed with ginger, scallions and soy-sauce it was cooked just enough so that its flesh was only just falling off the bones. One ritual that had to be observed was not to allow the fish to be turned over in stripping it of its fishy flesh. A local belief was that turning over your fish put your boat at risk of capsizing on the way home, so if you hoped to return to port safely, you dismembered your garoupa properly, stripping the backbone neatly away to expose the flesh below. The guest of honour was then served the jaw muscles in the fish's cheeks. All this was followed by Singapore noodles or hang-chow fried-rice to finish. Sweating and replete, we licked our sticky fingers and waved to our junk to come and pick us up; then happily cruised homewards in the dusk with younger children dozing on parents' laps, the bigger ones playing card-games or just lolling back reading their latest Nancy Drew or Roald Dahl. The lights on the Peak would beckon us home as the sun sank into a glorious sunset over the islands to the West. The weekend was over. It was back to work or school next day but that evening we were relaxed and delighted by the dramatic beauty of our new home: Hong Kong.

As the days and weeks went by, the weather cooled and the humid and hot Southerly Monsoon changed to the cooler and drier North Easterly

Monsoon which would dominate the weather until the following spring. There had been little rain during that summer when most of the annual rainfall should have occurred. The reservoirs were nearly empty and with little prospect of rain until next summer emergency rules to conserve water were in force. In our flat the mains water ran for only a few hours a day; if you didn't get your shower before 9pm, you went to bed unwashed. For the people living on the hillsides in the squatters' villages, the situation was particularly dire. Only one or two buckets of water a day for each household were allowed at the controlled stand-pipes.

The end of the summer Monsoon is marked by the Lantern Festival when we all eat Moon-Cakes to celebrate the harvest moon and we carry little paper lanterns alight with candles up to the Peak or round Victoria Park. Those with gardens or balconies with a view of the huge pink harvest-moon held parties. We all had the day off work to mark this very ancient Chinese festival. But in 1974 Lantern Festival passed without any rainfall, so the government was considering draconian water control measures when in early November Carmen arrived. She swept across the South China Sea having devastated the Philippines, packing big cyclonic winds amid her swirling rain bands. She scored a direct hit on Hong Kong. Huge ships were blown ashore, junks and sampans smashed to matchwood and many people were drowned. Low lying villages were flooded, power lines disrupted, huge trees blown down. People caught in the streets were struck by flying debris, bamboo scaffolding was blown awry on building sites. But still the rain came and came and came and the reservoirs were filled to the brim. Nowadays we rely less on our own reservoirs as water is brought down from China through the huge pipes that run alongside the railway lines near the border crossings and we have forgotten those days when water was precious and the Water Department's director[33], tongue in cheek, recommended 'Showering with a Friend'!

Typhoons are thrilling and dangerous events. Great rotating storms that start over the tropical seas East of the Philippines, they gather power as they travel Westward picking up water and heat energy. They are tracked nowadays by satellites, but back then reliance had to be placed on

---

[33] Tom Tomlinson; father of the editor of this book.

reports from ships at sea and weather stations on land; sometimes the forecast was a bit vague as the Royal Observatory was unable to plot the cyclone's exact centre. A series of signals was, and still is, displayed on signal masts and on the TV, nowadays on mobile-phone apps too. Number 1 if a cyclone is within 400 miles, number 3 as it gets closer and is bringing stronger winds; at this all small boats have to be back at their moorings. At number 8 it is blowing dogs off chains, the sea is wild and schools and shops close their doors, ferries stop running and double-decker buses hide in their sheds. A direct hit is number 10; the rain belts down, drains and nullahs overflow, mudslides block roads, trees are shredded. Nowadays very serious damage is unusual in Hong Kong, although the major storm of 2018 took months to clear up after, even with the help of the soldiers from the PLA garrison. Fatalities are nowadays very rare.

In the years before we came to Hong Kong, Typhoons Wanda of 1962 and Rose in 1971 had caused huge damage and many deaths. The heavy rain from Rose weakened several hillsides such that the following June, after several days of heavy Monsoon rain, a hillside in the residential area of the Mid-Levels suffered a massive landslip in which a block of luxury flats collapsed in a sea of mud. This was the Kotewall Road disaster which killed 67 people. Residents were trapped in their flats by fallen masonry and rivers of mud. Victims drowned in the mud or were crushed to death. It took several days to reach survivors in the unstable wreck and severely injured people had to lie immobile near the corpses of their family. One famous lawyer, who lost his wife that day, said he only kept his sanity while trapped in the dark because a radio in a nearby flat was still playing and he knew from the news bulletins that rescue was coming. On the same day. a squatter village in Sau Mau Ping also suffered a landslide killing another 71, mainly children, as well as several rescue workers.

These two tragedies resulted in profound changes. Firstly, building regulations were brought up to date and rigidly enforced. Secondly, the newly arrived Governor, Murray Maclehose, accelerated the enormous task of rehousing the hundreds of thousands of people living in unregulated squatter huts built on the steep hillsides. These could then be cleared. He inaugurated the Low-Cost-Housing Scheme, one of the Colony's great triumphs in social care that provided safe housing for the hordes of refugees who had arrived from China since 1949. There is still a huge scar on the hillside above Kotewall Road nearly 50 years later. No one has tried to build there and it is left to nature, occupied by wild boar, muntjac deer, snakes and other wild animals.

After Carmen blew away over the mainland[34], we enjoyed cooling breezes and sunshine. Our warm clothes came out and jackets and ties were worn again. Well-to-do tai-tais collected their fur coats from cold-storage at the Dairy Farm building and we threw off our summer torpor. Swimming-pools, previously thronged were now used only by keep-fit types doing laps, while the tennis-courts were fully booked. We went on energetic walks in the country parks, we climbed Sunset Peak on Lantau Island.

With Gareth on the wire; racing the 505.

---

[34] Mainland; the PRC as opposed to the Colony.

An old friend from sailing days in North Wales invited me to share a racing-dinghy with him. Gareth and I bought a second-hand 505 and joined the Yacht Club at Kellet Island in Causeway Bay. The 505 class had held a world championship the previous autumn in Hong Kong, sailing off the beach in Repulse Bay. For the year or so before that the class had been extremely popular, but many owners found the boats too difficult to manage. You need to be fairly athletic as well as skilful to handle this fast and powerful dinghy in anything more than a light breeze. The 505 could bite back if mis-handled. Inevitably there was a glut of boats on the market and consequently the prices were low.

Gareth and I alternated the roles of helmsman and crew. Both jobs were exacting and the crew had to simultaneously manage the overlarge spinnaker and swing from the trapeze wire to balance the boat. A capsize was the price for a mistake while learning to drive the 505, but the water was warm, if a bit dirty in places. One of my first outings in a 505 was across the Harbour towards the wreck of the old Cunard liner *Queen Elizabeth* that lay on her side, a burned-out hulk not far from the Yacht Club. Our little blue boat skimmed over the clear water, and looking down we could see the decks of the once noble ship far below in the depths. Sailing was to become my main relaxation and most Saturday afternoons Gareth and I turned out for the races in the Harbour.

As our first winter approached the sun shone while a cool breeze blew from the north. Local people stopped going to the beaches as it was now the time of the year for walking in the hills even though the sea was still warm and the sandy beaches deserted. We celebrated Christmas with an American-style butter-ball turkey for dinner, Joan made mince-pies and Diane was showered with presents from Bill and our other new friends. Of course, people went on falling ill or having babies but the brief change of pace was welcome.

A month or so later, Chinese New Year arrived with cold wind and rain. Coats and gloves were retrieved from the bottom of drawers where they had lain unworn for nearly a year. The dismal grey weather was matched by the greyness of downtown Hong Kong. For several days everything was closed; no cinemas, no restaurants, the shops boarded up. I had imagined

Lunar New Year as a festival with music and lion-dancing, incense burning and happy throngs of people celebrating under lanterns hanging in the streets. There weren't even fireworks because they had been banned during the Star Ferry Riots a few years earlier when gunpowder from firecrackers had been used to make bombs. In fact, the place was deserted, everyone who had not headed for their home villages in China would be locked away at home with their families. All the amahs disappeared too and without help in the home, many expats headed for the hotels and beaches of SE Asia to escape the weather and the temporary tedium of our ghost town; quite apart from having to do their own cleaning and cooking.

Shortly after Chinese New Year, Vivienne's cousin was to be married and since these celebrations are always large ones we were invited too. A few expatriates at the event seemed to add to the family's prestige in some way. We arrived at the Chinese restaurant in good time and found ourselves seated with some other gweilos. A lot of Shao-Shing rice-wine seemed to be consumed though we, like many others, stuck to warm orange-squash. The meal of many courses was delicious. Speeches and toasts followed, the bride looked delightful in her crimson cheong-sam with a slit right up one thigh and the groom seemed terribly young. As things began to wind down, oranges were served to every table. I noticed that only we gweilos were still sitting and chatting at the table, peeling our fruit. The rest of the guests had pocketed their oranges and scooted off home. The arrival of oranges means 'that's it, off you go!' They are a symbol of good luck and fortune, being round and golden; only a foreigner would think they were to eat!

The anniversary of our arrival in the Colony came and went and our feeling of home and of belonging increased. I was still writing long letters home to my father and to close friends describing our new way of life which still felt novel and worthy of description and comparison with how we had lived back 'home' in Britain. Nevertheless, the Hong Kong way was becoming more and more the usual way, and English values and English practices seemed less relevant and perhaps narrow and antiquated. Hong Kong seemed to embrace change, applauded innovation and its economy was growing by leaps and bounds. I started reading the financial pages of the local paper although I had no spare money to invest. I was excited by how the local GDP was rising fast, the stock exchanges were booming but I was a bit frightened by the inflation of property prices.

Sascha, our second daughter, was born with Bill in attendance at the Matilda Hospital during the autumn just after we had moved from our first tiny flat in a garden to a bigger but scruffier apartment nearby, but still in Shouson Hill. Joan's mother flew out to help and we held a party at the new flat after the christening at St. John's Cathedral. Bill made a speech that linked Sascha's arrival with our family putting down roots and becoming part of the local community. We had acquired a fluffy white dog from the RSPCA, Spitz, who was of a very independent nature and liked an outdoor life wandering round the nearby apartment blocks getting to know everyone. Diane was now attending a nursery group in a nearby garden-flat at Cheerful Villas, which belied its name, being a grim looking grey housing development a few hundred yards from us. She walked down leafy Shouson Hill Road accompanied by Joan and Spitz every morning.

Diane and Spitz.

Sascha's christening at St. John's cathedral.

Of course, Hong Kong is celebrated for its shopping and its restaurants. Still on a small salary, our shopping was quite restricted, but to mark our wedding anniversary I visited the Central Building, now long ago replaced by the massive Landmark complex, to visit Sennet Freres, a Parisian jeweller, and buy a silver necklace for Joan. It was of a somewhat avant-garde in design in 1975 and she still wears it from time to time. My father was remarrying[35] that autumn back in England so Joan and I, unable to get to the wedding, went to an antique shop in Hollywood Road near the Man Mo Temple and bought him a pair of large blue porcelain elephants as gifts.

Mothercare, who had recently opened their first Asian store in Central, sold us a McLaren stroller to push Sascha around as soon as she could sit properly. Chinese babies were hardly ever seen riding in prams, as their mothers preferred to carry them and we often saw babies in slings on their mothers' backs as they worked in market stalls or paddled their sampans using a single long oar slung from the stern. Babies from better-off families were carried by their amahs. In the street, Sascha was a magnet with her very blonde hair that passers-by liked to stroke, presumably for luck. She hated this of course and would strike back. One day a blind beggar thrust

---

[35] My dear mother Mary had succumbed to cancer sometime earlier while I was still living in England.

127

his bowl in front of her as we passed and she grabbed a fistful of coins from it. The poor man was so shocked that it took a while to calm him down.

We started to find restaurants that we loved. We couldn't afford the very expensive Chinese and European restaurants, and anyway we preferred the delicious food we could enjoy in less luxurious places. At the Java Rijsttafel in Kowloon, we went for Indonesian salads and curries, but sadly they closed down as rents rose in the 1980s. At the Gaylord[36] at the end of Nathan Road near the Harbour we ate curries while Sascha lay in her basket under the table gumming on Naan bread. Jimmy's Kitchen was then in its heyday with a vast menu of choices but whatever else we would choose; meals were always completed with Cherries Jubilee flambeed with a fiery flourish at the table. On Sundays, the Repulse Bay Hotel served a buffet lunch for only $22 and we piled our plates high with such delights as smoked salmon, which in those days was an expensive and rare luxury. In the winter cold months at the hotel there were Swiss fondue evenings in the wood panelled library, occasional conflagrations just adding to the fun.

Then there was the American Restaurant in Lockhart Road, surrounded by girlie bars and hotels letting rooms by the hour. The food was Pekinese and certainly not from the USA. For thirty or forty years, we went regularly until they too closed their doors. The same waiters had served us from the same menu for all that time: chilli prawns, deep fried seaweed with bamboo shoots, dried shredded beef which you stuffed into sesame pockets, fried rice and then to finish off, apples dipped in toffee which cracked when dunked in iced water before you bit into them. These were our favourite restaurants and now they are no more. On special days out in the New Territories, we went to the Lung Wah where we gorged on sweet roasted whole pigeons; they were reputed to slaughter a thousand birds a day, or at the Shatin Heights looking down on riverside fields while eating grilled satays on little sticks of bamboo and where nowadays sits a modern city. With food like this we didn't need expensive wines as the local beer, San Miguel, was perfect.

---

[36] The Gaylord after several reincarnations is still open, now in Peking Road serving great curries.

CHAPTER 17. INTERLUDE: CLOUD IX

On *Cloud IX*, it was a filthy night and the yacht was tearing along in a following sea. Christer our Swedish skipper was below poring over a soggy wet chart; seawater was sluicing down the companionway with every passing wave and adding to the damp chaos in the cabin. Our next turning mark in the race was a rock with a light tower. It stood in the wide south facing bay of Cheung Chau Island, now a few miles downwind of us. *Cloud IX* was a very wet boat, long and narrow and rather old fashioned compared with most of the others in the fleet, but she needed only a small rig and sails which were light to handle for a small crew. We had taken the spinnaker down a few hours earlier as a fresh north-east monsoon surge arrived with a thirty or forty knot blast of cold wind. The temperature had dropped ten degrees and rain poured down. Discretion was the better part of valour we had decided and I doubt the boat would have gone any faster with a larger press of sail.

We couldn't see much in the driving rain although ahead of us was the yacht *Ciel* which had lost control of her spinnaker now streaming forward from her masthead, the sail and sheets, accidentally released, were now thrashing around in the gale at the masthead 50 feet above her deck. We could see that someone was courageously climbing their mast to secure a rope onto the head of the sail and haul it back down. *Ciel* kept disappearing in the murk as bands of rain swept between us. The light tower we sought was invisible in the rough sea and we were relying on Christer's calculated position; GPS wouldn't be invented for another 20 years. If we got it wrong, we could end up hitting the rock or being swept into the shallow bay, was this a gamble we should take?

Christer emerged pulling on his sou'wester cap and looking like a Baltic fisherman with his bushy moustache. 'Too risky' was the decision, so we opted to change course and sail right round the whole island before turning towards the finish line upwind, back towards the east. I was shivering at the wheel and Christer noticed I was chilled. 'Get forward and hank on the smaller jib for the next leg! This genoa is too big'. I hauled the small sail in its bag up to the foredeck. The bow wave cascading over the deck felt warm compared with the icy wind. After twenty minutes wrestling down the genoa and hoisting the small jib in its place I was warm again and sweating inside my waterproofs with the all effort.

Our caution didn't win us any trophies but we felt relaxed and happy as *Cloud IX* crossed the finish line and we headed back to our mooring at the Yacht Club. The Sunday dawn was just breaking and, I knew that in an hour or two I would be back at home in bed. I had had a few hours away from the practice, I felt tried by the elements; we had been in a little bit of danger and we had made no mistakes.

We left the boat as it was, sails bundled up anyhow and ropes lying uncoiled on the decks, wet cushions in the cabin, galley sink full of dirty plates and coffee cups, and empty beer cans in the bin. Dai Mui would clear it up. Dai Mui (Big Sister), was our boat-boy (male or female they were all boat-boys) and her job was to keep the boat clean and tidy, fold the sails and coil the ropes, she scrubbed weed off the waterline, she rinsed the salt off the equipment and brushed and oiled the teak deck. She put up the sun awnings to protect the woodwork and took them down again before we came aboard. When storms threatened, she secured everything and doubled up on the mooring lines. She was always there in her little sampan to help with the mooring lines when we sailed in or out. Dai Mui was the leader of the boat-boy 'mafia' at the Yacht Club. Even if you thought you didn't need a boat-boy you had reckoned without the mafia, it was almost a protection racket. When years later I bought a yacht myself, the excellent Dai Mui looked after it, she was not expensive but you had to have her or one of her colleagues to look after your boat.

There was an overnight race for the cruiser fleet every few weeks in those days, starting on Saturday afternoon and finishing on Sunday, even the small boats were back at their moorings by lunchtime. The courses made use of the different islands and rocks around the Colony as turning marks. In the north-east was Gau-Tau, a lonely rock off the Chinese coast with a flashing light tower, to the east was Wag Lan, a steep rock with a proper lighthouse upon it, visible for tens of miles out to sea. Po Toi and Castle rock stood off Stanley and Repulse Bay and out to the west were the Soko Islands on the way to Macau. Skippers were always in need of crew and we dinghy-sailors were eagerly sought after and pampered with picnics and cold beer when aboard.

Night Sailing

There was little sleep as we trimmed and steered, changing between sails and setting the spinnaker. Accurate navigation was needed to make the shortest course and to avoid the few dangers we might run into; the skipper usually did this. Some of the more competition-minded skippers had their crews sitting up on the weather deck all night to help balance the boat with their weight.

At Lunar New Year we raced the forty miles or so to Macau, where we stayed for a couple of nights before racing back home. Macau in those days was a bit of a sleepy backwater, not like the flashy rival to Las Vegas it is now. There were little Portuguese restaurants with checked tablecloths, we ate Chinese seafood at Formica tables on jetties over the sea. We drank Vinho Verde and Dao wines, went to the Military Club for African Chicken, chewed on chorizo at cafes on the roadside. In the street we had to dodge the local revellers and their exploding firecrackers, the pavements were knee-deep in the resulting debris of red paper.

Before sailing back to Hong Kong, we loaded boxes of cheap wines aboard our boats, hiding them under the floor boards to be smuggled back home where wine was subject to duty. Though why we worried about this subterfuge I can't think, for nobody ever checked the boats when were back at our moorings.

To go sailing you did not need your own boat, you just had to be good company, know the rudiments of sailing and not be sick on the skipper.

CHAPTER 18. ASTHMA AND OTHER AILMENTS.

In the 1970s, asthma was less common than today. Great advances were being made in its management and a number of drugs had recently been introduced that were game-changing. The rather toxic Isoprenaline inhalations had been replaced by Ventolin administered from an aerosol that patients carried around in their pockets; inhaled steroids like the Becotide aerosol were used as a long-term preventative instead of the old weekly injections of ACTH. Intal was still popular although I suspected that this white powder, inhaled through a special whizzer device, was more a placebo than anything else. At least, just like homeopathy, it had no side-effects. The recently discovered house-dust mite was ruthlessly hunted down and exterminated.

These advances had made life much better for our patients and the child who had previously sat wheezing on the touchline could now join his friends on the football field. Still, we had many cases needing hospital admission. I taught parents to monitor breathing performance at home with a simple meter and adjust dosages by the results. I had to overcome arguments about not needing his drugs when he was well; prevention had to be seen as paramount. Action was the byword, passivity on behalf of parents or physician was no longer acceptable; asthma was controllable.

An acute asthma attack is a distressing thing to witness and much worse to experience. The child is pale, lips are blue, his whole effort is focussed on moving air in and out of his lungs; to exhale this breath then gasp in the next. He cannot talk, he can't take a drink. Every muscle, every nerve cell is concentrated on breathing. He is anxious and he is exhausted. Left in this condition he may die.

At the time we were becoming more aware of the biochemistry of acute asthma with the introduction of rapid methods of analysing the levels of Oxygen and Carbon Dioxide in the blood. During my five years of training blood-gas measurement had moved out of the research lab and into everyday practice and what we learned from this came as a shock. Many of our previous interventions actually made the patient worse. Sedation had been regarded as important to allay anxiety and calm down the breathing, it certainly sedated the patients who seemed better and their breathing easier, however when we checked their blood-gases, we found the Carbon Dioxide had gone up and the Oxygen down; the opposite of what was needed. Aminophylline reduced wheezing but lowered the blood Oxygen level too. We had previously withheld liquids, worrying that we might strain the heart and flood the lungs with fluid, that too was proved wrong and in fact they needed a lot of IV fluid to correct dehydration. To my horror some of the doctors were still using the old methods which were out of date and dangerous.

I kept to the scientific approach and involved the lab. We gave a lot of IV fluids which alone relieved the breathing and we administered high levels of Oxygen usually in a tent. The main drug we used was a steroid in a big dose; it did take several hours before it was effective but it was the mainstay. Until it began to work, we had to use Ventolin and old-fashioned Aminophylline despite their harmful effect on Oxygen levels. This was all very labour intensive but we were able to shorten hospital stays. Through the parents' grapevine the word began to spread and we added many more youngsters with asthma to our list, all of whom were taking regular preventive medication and all regularly tested their breathing and reported to the clinic. Sports were encouraged and less schooldays were lost to chesty coughs.

Croup was another common problem among toddlers. The intensely dry cold air of the early part of the year made the little ones with bad colds cough with that characteristic metallic note. All of a sudden, usually in the night, they would have breathing difficulties. I spent many evenings sitting with these little mites in steaming hot bathrooms as they fought for breath. The hot humid air gradually eased their breathing. Often admission to hospital was needed if improvement didn't come quickly. Nowadays an injection of a steroid brings relief in a few hours but in the 1970s we were perhaps too concerned about possible side effects and really didn't like to give such a powerful drug to toddlers.

My nights were frequently disturbed by calls from patients and I would have to drive to see patients all over the island. I have lasting memories of wild rainy nights on my way to yet another case; driving along the coastal road towards Stanley Village. I loved the brief sight through the murk of the flash of the Wag Lan lighthouse out on its lonely and windswept rock. Miscarriages, toothache, difficulty in breathing, old men with urinary retention, heart attacks; often I could sort these out in their homes without needing to send them to hospital. Patients with uncomplicated coronary thrombosis in the early 1970s were routinely cared for in their own beds at home.

A typical call had me hurrying to a high-rise block in Repulse Bay where Ed was having breathing trouble. An American and the local agent for a manufacturer of fountain pens and ballpoints he was a heavily built man in his 60s. He was sitting up in bed gasping for breath, his lips were blue and his pulse racing. My stethoscope told me his lungs were waterlogged and he was in heart failure. After I had given him a diuretic injection I called for an ambulance; Ed was in serious trouble and needed oxygen at the very least.

The ambulance crew brought up an oxygen cylinder and Ed's breathing improved marginally but we needed to get him down to the ground floor and into the ambulance. He was carried in a chair to the tiny lift which would only accept me and one ambulance man as well as Ed. Half way down he arrested; he stopped breathing and had no pulse. We stopped the lift and lay him on the landing floor to give him cardiac massage and the 'kiss of life' for which we used a special tube that we all carried in our bags in those days. We were not to be alone for many minutes as the door of a flat opened and a very angry resident rudely told us to leave or he'd call the police; we were making too much noise. Ed started breathing again and we continued on our way to the hospital. He recovered but decided to retire and return to Wisconsin.

Bill was doing more and more maternity work so I was spending much more time in the hospitals. While waiting for the deliveries to progress I would visit my patients on the other wards where I began to meet more and more medical colleagues. There were two big practice groups: Vios and Andersons. Vios was the smaller, it had been founded by an Italian doctor just after WW2. Andersons however dated back much longer to the early days of the Colony when the great Patrick Manson set up his practice on the Island. Manson is regarded as the father of tropical medicine and his research on the transmission of disease by insect vectors led directly to Ronald Ross later demonstrating that malaria was spread by mosquitos. The Manson practice went through several doctors' hands and became Anderson and Partners sometime between the wars. They had been responsible for controlling several epidemics over the years and they owned a motor launch to do quarantine inspections of arriving ships as there was still a real risk of importing Smallpox and Plague.

These two practice groups were quite self-contained and had their own specialists, so apart from maintaining friendly contact we seldom worked together and of course professionally we were in competition with them. There were many other private doctors, mostly locally trained, who worked single-handed. They would sometimes ask me for help or to stand in for them if they needed time off. I was already starting to treat some Chinese speaking patients, which I took as a great honour and an even greater responsibility. I often wondered why, with a wide choice of doctors with whom they could communicate easily in Cantonese, they chose to consult me.

A young couple visited me and refusing a translator told me in laboured English of their hopes of having a baby. They had been married for two years and pregnancy had eluded them. I started thinking of sperm-counts and getting a gynaecologist involved, but they stopped me; they had seen a gynaecologist already. The reason for their plight was essentially non-consummation because intercourse had proved too painful for her and the situation was becoming desperate. She had come to hate even the idea of sex and her vagina would be dry and sore after trying. The prospect of giving sex-counselling to this couple across the language divide was daunting, even insurmountable, as I was ashamed to acknowledge. I did my best and gave a perfunctory and rather desperate description of foreplay and so on. I ended up handing them a virgin tube of KY lubricating-jelly from the examination trolley beside the couch and suggested they use it during foreplay. Too embarrassed at my ineptitude and even cowardice I would not even charge them a fee. The story has a happy ending though; sometime later a Chinese couple arrived in my room with a young baby, I did not recognise them I'm afraid, but it was the same couple full of gratitude for 'giving them a baby boy'. They wanted me to join with them briefly and see their miracle. It was so often like this; the patients you think you've done the least for are sometimes the most grateful.

Heinrich was very tall young Dutchman who worked for a large European trading-company and travelled to source bulk materials from all over SE Asia. He spent a lot of time in Indonesia, Thailand and the Philippines and was usually away for weeks on end in fairly primitive areas. Agricultural and mineral products had to be sought where they were made. He was a good-looking man with a jaunty confident manner and a Zapata moustache, so fashionable at the time. He was obviously popular with the fair sex as my clinic was often his first port of call on his return to Hong Kong suffering from one or other of the diseases associated with the enjoyment of the delights of Venus and Eros. After taking the necessary swabs for culturing in the lab. I would give him an antibiotic injection. Every time I did this he fainted. Apparently, he had always fainted as a schoolboy in Holland when getting his vaccinations. He was now a very big man and was a heavy weight when he crumpled over and pinned me to the wall. 'Please lie down for your shot' I would implore. 'No', he would say, 'I promise not to collapse'; but he always did. After a year or two Heinrich changed jobs and got married. Then I started looking after his children instead of him. I don't recall whether they would faint after their shots too.

Venereal disease was a common problem for men who had to spend a lot of time away from their families while on business and some of them consulted me when they managed to contract an unwanted infection when away from Hong Kong. Attitudes to sex in some Asian cultures were very permissive; girlie bars and massage parlours attracted lonely and randy men and my patients were no less tempted than most others. People spoke of 'Kai Tak Rules', named after the old airport in Kowloon: what happened outside Hong Kong stayed outside Hong Kong. Unsportingly the diseases caught outside Hong Kong did not obey these rules and men sometimes found themselves developing symptoms several days after getting home and having been lovingly reunited with their wives. The knots they tied themselves into, trying to push their spouses to a doctor for a check without admitting their own predicament were worthy of a chapter of their own. My advice was to make a clean breast of it and ensure that effective treatment was not compromised; otherwise if the truth emerged later, they would find it difficult to ever recover their wives' good opinion and trust.

There was one local doctor who was famous for his interest in venereology and also for self-advertisement; in fact he could be a most amusing man altogether. But one time I thought he went too far when he wrote a magazine article recommending that husbands who had erred and been infected should go and get treated themselves without informing the wife. After the all-clear they should resume their relationship and then if, and he advised only if, they themselves developed their symptoms again then they would they know that they had indeed passed the infection on earlier and would have to own up. I was very shocked reading this unethical nonsense and considered reporting the matter to the Medical Council. Fortunately, Bill advised me to mind my own business and not involve myself with this clown of a man. Perhaps it was just meant as humour.

It was, in fact, not only men who were troubled by the diseases of love. A nice and likeable Australian lady who had recently become engaged to the widowed father of two boys who were patients of our practice requested a VD check-up, having enjoyed a somewhat chequered past as a member of a hippie group in her previous life. I carried out a thorough check and she tested positive for Syphilis, a disease that was quite rare but nevertheless still a relentless and insidious enemy. Since she had been with her new boy-friend for several months he too needed to be checked but she surprised me by refusing to inform him. 'Then I must tell him myself' I told her.

'What about patient confidentiality?' she asked. I explained that the family had been in my care for a while and my first duty clearly lay in keeping them safe and I was quite prepared to take the consequences of speaking out. She then did the right thing and I was able to test her fiancé a few times over the next year or two and he remained uninfected, I treated her too and she was cured. The couple did not marry in the end but were happily together for several years.

While this was going on we were seeing the arrival of expatriates from several nations who had left Saigon as the US army withdrew gradually from the war there. Several career diplomats had arrived and a local consulate had rented flats for them in Repulse Bay. One of these men arrived in the clinic asking for sleeping pills. I could imagine the stress of the escape from a war-zone and was happy to give him a few days' supply, but it turned out to be an entirely different matter. His wife was a Vietnamese lady and she had become suspicious of his friendship with another woman in their block of apartments. A few nights earlier she had run, screaming and angry into their bedroom armed with a long kitchen knife which she then plunged deep into the mattress right between his legs. Though unscathed he found he couldn't sleep since the incident; women from that area of Asia have always had something of a reputation for mutilating errant husbands and perhaps he was right to be worried

Some extremely rich people live In Hong Kong and I was beginning to see some of them as patients. Many lived unostentatiously and were only revealed as billionaires when occasionally their names appeared in the business section of the South China Morning Post but others paraded their Rolls-Royces and flashy jewellery in the society columns. When they came to consult me, they would be dropped outside by their chauffeurs and ascend to the clinic bedecked in their diamonds and Diors, sometimes even accompanied by a body-guard. One well known society lady rushed unannounced into the waiting room in a panic on a busy morning with her toddler son. She had been driving through Central when little Dominic had leant over and grabbed one of her solitaire earrings and promptly swallowed it. I wasn't sure whether her concern was for Dominic or the solitaire but a quick X-ray showed a diamond of several carats that had safely passed into the stomach and fortunately had not got stuck anywhere higher up. I assured her that it should now pass unobstructed in the next few days but to come back if it did not appear. It did not and she returned. Dominic was subjected to another X-ray; nothing! There was no diamond to be seen. She grabbed the still wet X-ray film and dashed out yelling 'The bastards, the bastards!'. Clearly the servants back at home on the Peak were going to be in trouble. Had they missed finding the precious stone in Dominic's diapers or perhaps they had stolen it? I never found out.

I was becoming known and patients and sometimes strangers now stopped me in the street or the supermarket for a chat or to tell me their symptoms; after all, free medical advice is a rare commodity and should be seized when it may be. Walking one lunch-time past the Cricket Club, in those days on an open space in front of the China Bank, I encountered Mrs Patel with her eldest boy, she chatted away complaining of how long it was taking for one of the boy's milk-teeth to fall out. As she talked and laughed, I felt for the loose tooth and gently removed it from the boy's mouth. He just grinned at me as I turned away from them both and without further comment placed the offending incisor in his mum's palm.

CHAPTER 19. PARTNERSHIP.

Once my first twelve months were up, Bill had asked me to join him as partner and so as the practice prospered and grew so did my income increase. Joan and I could start to save some money. Joan found a young Filipina woman to live in and help in the house and with the children. Her name was Faye. It sounds terribly grand to have a servant but in reality, running a household in the tropics made some help necessary, particularly when I was always being called away by patients. Faye had her own room near the kitchen and she was a happy presence in the household; the children loved her and when they were scolded by their parents would go for comfort to Faye. Mealtime discipline went completely to the dogs. Threats that they would go hungry if they did not finish their greens meant nothing if later, they could sneak into Faye's room and fill up with some of her lovely soup-noodles.

In those days domestic-helpers were all called amahs, though that term really implied the helper was Chinese. Nowadays helpers come mainly from SE Asian countries.

We have always had a helper living with us and keeping house; they usually stay for a few years and then move on with tears shed by one and all as we parted. The children generally loved them dearly and they often came on holidays abroad with us. We had now been in Hong Kong for three years without a proper break so one summer afternoon found us at Kai Tak catching a flight to England for a month away. We enjoyed a busy holiday visiting friends and family all over the North West of England and even managed a few days in France. Sascha, now two, had never been outside Hong Kong and found England a strange place, especially the food. She missed her noodles and rice, so when we were in a Liverpool Chinese restaurant it was difficult to persuade her to leave as she gorged on bowl after bowl of spicy noodles. Finally dragged from the table she stuffed handfuls from the serving dish into her pockets. So much for her grandmother's fine cooking!

Then back home to the heat and humidity. Joan feigned surprise that in my absence heaps of corpses had not built up on the roadsides round

Repulse Bay. I had to ignore jet-lag as the phone was already ringing even before we had unpacked. Joan rolled her eyes as I ran for the door, car keys jingling, off to the Matilda; a new baby was arriving and they wanted me there.

The weeks away gave me a chance to reflect and to compare my practice and lifestyle with those of my friends still working in the NHS. Private practice in Hong Kong seemed to be much more fun, perhaps because it was less circumscribed. We had immediate access to specialists and to investigations such as CAT scanning with no waiting lists. True, the patients had to pay for this but for the type of people that I cared for in private practice the convenience was worth paying for, and it was usually the employer or an insurance company that picked up the bill. At that time the Hong Kong government's free hospital service was inferior to the NHS but most of its users never paid a cent because income tax at only 15% was only levied on high earners, there was no VAT and no National Insurance to pay either. I had to admit that I had given up the security, the 'iron rice-bowl' was the local expression, of a monthly salary and eventually a guaranteed pension for the somewhat precarious finances of our practice and indeed sometimes I did lie awake in the small hours of the night worrying over the next bills for rent or payroll.

Locally in private practice the prevailing philosophy was that patients' problems should be resolved as quickly as possible. There was no point in putting off today's problems until tomorrow when you would still have to spend the same amount of time solving them. So, we tended to keep the clinic open until we had cleared the day's backlog of requests for appointments. The X-ray labs and the specialists took the same approach and fitted in urgent bookings where they could. Cynically, one might say that we were just worried the patient might go elsewhere and we would miss out on our fees; of course, this was a factor but we really were anxious to help; to relieve pain and anxiety and get people back to their normal lives. Somehow in my old life in the NHS I seldom felt the need for urgency except in dealing with emergencies, and it seemed acceptable to let most patients wait, after all the word 'patient' is derived from the French verb for to wait. Here in the frenetic world of Hong Kong nobody wanted to wait; neither patient nor doctor.

Still more new patients were arriving as developers opened up new apartment blocks in Repulse Bay and nearby Stanley. Bill and I worried about raising our fees as inflation was eroding our working profit. We bit the bullet and raised our charges. A few patients mentioned it but the numbers continued to rise.

Teaching is an important part of practice; it is often and truly said that students teach you more than you teach them. I had been appointed an honorary lecturer at HKU soon after my arrival in the Colony and made a paediatric ward-round at Queen Mary Hospital with a small group of students every week. We walked round the beds and cots saying hello to the children and discussing the problems they presented. In Hong Kong the medical students are those that have striven the hardest in the hot competition that starts at age 4 when some of the better kindergartens even had admission exams. For their children to succeed most parents expected to pay for extra tuition after school. Sports were discouraged and all effort and time were devoted to getting excellent exam results. The result probably was, and still is, that the top students were rather narrow in their experience of life and valued book-learning very highly. This is worrying in a clinical situation where observation, intuition and logical thought are needed; these three things are of little value without the book-knowledge but they do need to be there as well. It was my aim to get my bookish young men and women to look and think on their feet and to understand that often the answer was not to be found in print. They were brilliantly informed of facts and up-to-date on recent developments in a way that most students in the UK were not, but they found difficulty in actually managing a patient and his disease. I spent my ward rounds teaching clinical-method; just examining the patient needed emphasising; feeling the abdomen for an enlarged spleen or locating tender spots, eliciting tendon reflexes; sometimes just standing back and taking a careful look. All the things they would be doing for the next forty years. Properly done these low-tech methods save time, save doing unnecessary tests and save money. They probably save lives too.

All these students would qualify as doctors within 2 or 3 years and become colleagues. It was a great pleasure to meet them years later in the wards or at conferences. Many years after I had given up teaching at Queen

Mary, I encountered some bureaucratic difficulties over my specialist accreditation and had to be interviewed by a clinical Professor at the Chinese University. We immediately recognised one another; he had been one of my students decades before. After some reminiscing, he was able to cut through the red-tape and solve my problem.

Bill and I were spending a lot of time in the operating theatre, because not only was a paediatrician needed to attend at caesarean-section deliveries but he liked me to be his assistant in general surgery too. Bill was one of those surgeons with the knack of making everything look simple. When he opened a belly, somehow the tissues and organs were exposed looking like an anatomical textbook. So many other surgeons I had assisted over the years made the operation field look like a complete dog's-dinner and I wondered how they coped.

I was always there for caesareans and, before the baby appeared, was usually left to look after the father. The attendance of fathers at deliveries was quite a new thing and they were often in a nervous state. First though, they had to change into theatre scrubs, masks and caps and some did not seem to know how to manage even that. Others were full of confidence and were in everyone's way setting up their camcorders to record the best view of everything. Since all the rest of the staff had their hands full with getting on with the business of delivering the baby, it fell to me to control these wretched men; some fainting at the sight of blood while others revelled in watching what was happening in gory detail. One evening Bill was cutting his way through some thick fibrous tissue resulting from a previous operation. As the scissors snipped noisily away through the scarring the father, who was very bald, remarked it was rather being at the hairdressers. Bill, remembering the man's shining pate said that he was surprised that he could still recall what happened at the barber.

Sascha when she was about three years-old developed a great interest in death and what happened to people when they died. In an attempt to answer some of her questions I took her to the historic Colonial Cemetery in Happy Valley and showed her all the ancient tombstones dating back to the birth of the Colony. Many of the earliest belonged to young servicemen who had died in battles with pirates or simply succumbed to tropical diseases, their elaborate tombs were often subscribed by their mess-mates. As we were leaving this historic tree lined hillside, we passed a shining black marble slab engraved with the name J.R.James. Jimmy James had been a patient of mine and had died in his late nineties and was, I think, the very last person to be buried in the old graveyard. That evening as we entertained some new friends to dinner, Sascha, on her way to bed, came to say goodnight to our guests and spoke of her day out; 'I went with Daddy to the cemetery this morning and we saw all his dead patients.' A fine recommendation indeed to potential newcomers to my practice.

CHAPTER 20. CHINA OPENS UP.

Mao Zedong died in 1976 and the old Bank of China Building in Central had been covered with a huge poster commemorating his passing. A more liberal regime was emerging led by Deng Xiaoping who famously said that 'to be rich is to be glorious!'. China began to do business again with the rest of the world. Hong Kong was the natural stepping-stone for this new commercial opportunity and international businesses were moving staff and their families into the Colony. The practice was looking after many of these newcomers. One large American engineering firm had been quietly selling heavy machinery in the PRC for some years and was now seeing a huge increase in business. One of their sales staff, a newly arrived Welshman, David, was spending a lot of time away in China, weeks on end in fact. His problem was that there was great demand for western products to get China's businesses moving but potential buyers lacked cash and needed to pay in kind. David realised that he must accept bulk-goods instead and his first deal involved selling some machines in return for several tons of frozen shrimp. For these he found an Argentinian buyer also without currency; would David accept several thousand coffins? He found a chain of undertakers in the USA who wanted coffins and so the deal was struck. This opened up his business and soon he was travelling widely in China trading commodities as much as selling mechanical equipment.

David came in for a typhoid vaccination one morning and I asked him what the Chinese characters embroidered on his shirt cuff read. 'Oh' he said, 'You know, I don't really like rich food and every time I do a deal my hosts put on a lavish Chinese dinner and it upsets my digestion. So I call over a waiter and show him my cuff; it says 'Can I have some plain rice?''. David became a firm friend and we kept in touch for many years after he left Hong Kong.

Some patients did not cope in China as well as David did and they found the pressure of working alone far from home and virtually incommunicado too much for them. The human resources manager for a large local firm called me about a senior man who was alone in far-flung Xian in the final negotiations for a large financial deal with a PRC government undertaking. All his phone calls and faxes were being monitored by the other side and he was having to make multi-million-dollar decisions on his own. He was now having pains in his chest and the company was very concerned that he might have heart trouble. There was obviously only one course; postpone the business and hurry him back to Hong Kong. As a precaution I admitted him to hospital, but all the tests showed his heart and digestion were fine, he had unsurprisingly been over-stressed. In association with a psychiatrist colleague, we set up a programme for senior executives with two parts. Firstly, they had a thorough check-up including stress ECG in order to convince them of their robust health, then they were given counselling on stress management techniques. This was quite a success and several other companies enrolled staff members in the programme. Truly, in those days China was a wild frontier.

When a businessman was travelling, and at the time it was almost always a man, his family had to manage without him. This put a great strain on his wife and occasionally she would not stand the test. Alcoholism and extra-marital affairs were common enough but we also saw mental illness as a result. I was summoned at first light one summer day by Gaynor, the 12 years-old daughter of an absent father. He had been in China for a week or so and Gaynor became alarmed that her mother was acting strangely. She was running round the flat naked, packing suitcases with cushions, light bulbs, kitchen utensils and such useful items. She had her passport and was going to Kai Tak as soon as she could get a taxi. When I arrived at their first floor flat in Jardine's Lookout there was pandemonium. Gaynor's mother, Sadie, was indeed rushing around wearing absolutely nothing and the suitcases were now by the door ready to go. Sadie was absolutely manic, moving about picking things up and putting them down, mumbling to herself and laughing. Gaynor had unplugged and hidden the phone, she had been unable to reach her father and so had called the doctor.

I tried to persuade Sadie to put on her dressing-gown but with a cry of 'I know you Dr Howard!' she ran out of the flat and down into the car-park where the security-guard sat in his deck-chair and calmly watched the gweilos chasing around in the early dawn light. Finally, I had her cornered and forced her into my car, still she was undressed. I decided that I would have to take her myself to the psychiatric ward at the Matilda Hospital on the Peak. I was all too well aware that abducting any patient against her will was quite illegal and that I should seek certification from a magistrate first, but this was five o'clock in the morning and who knew what Sadie might get up to before the courts were open?

So off we went. I drove carefully, hardly able to change gears as my left hand was occupied in grasping Sadie's wrists to prevent her jumping from the moving car. We drove onto Stubbs Road leading up to the Peak. But, horror! As I rounded a bend there was a road-block. A young policeman in khaki shorts stepped out and waved us down. I stopped the car and as I wound down my window all my future life flashed before me; the headlines in the SCMP, the court-case, the Medical Council hearing to strike me off, the disgrace. Ruin stared me in the face.

He stared at the naked woman screaming and struggling in the passenger-seat biting at my hand with which I was restraining her wrists. He looked at me, looked back at Sadie then stood back and waved me on. Clearly, violent abductions were not a top priority that day! I managed to deliver Sadie safely to the hospital and before long she was returning to sanity under the care of the psychiatrists. She did forgive me, or maybe she forgot the incident. Gaynor never did forget and we kept in touch for years afterwards. Sadie, still a bit dotty, died in her 80s in California.

Not all these incidents had happy endings. A new patient came into the clinic in Repulse Bay and asked me to tell her how she could tell if her husband was trying to poison her. I considered that the most likely explanation for this question was not an uxoricidal husband but a paranoid-schizophrenic wife and I was accordingly extremely worried about her. I gave her a single Valium tablet, hoping it might calm her down temporarily, and she agreed to be referred urgently to a psychiatrist. We called a taxi and sent her off to the hospital. A few hours later the psychiatrist rang to say that she had not arrived yet, but by that time, as we heard later, she had set herself on fire and jumped to her death from a high window in her housing block. For a long time afterwards I had an 'if only....' feeling about her. Should I have done more and gone with her in the taxi or perhaps sent my nurse? Such questions are unanswerable but haunt you all your life.

As time passed Joan and I began to feel more settled and realised we were likely to be in Hong Kong for many more years. We decided therefore that we really should try and learn to speak Chinese. Mr Liu came twice a week to the clinic after the patients and staff had left, to give Joan and I lessons in Cantonese conversation. I think old Mr Liu learnt more about the fascinating details of the expatriate life we led than we learned Chinese from him, and after only a year or so I abandoned these sessions, though Joan started going to full time classes at a college in Kowloon Tong and became quite fluent in 'market' Cantonese and able to rattle away to bemused stall holders and understand by how much she was being swindled as she haggled with the fish-lady over a catty of live prawns or several still struggling pomfrets. She learned to sing the names of all the different varieties of local cabbages to the tune of Happy Birthday and can still do this if asked.

My language education was later much narrowed down and supervised by Vivienne so eventually I could ask my non-English-speaking patients to undress, sit up, lie down, take deep breaths and so on. And that was about as far as I ever went with Cantonese I'm ashamed to say.

Joan and I visiting a farm near Canton.

The opening up of China started for us personally in 1979 when the Red Guards were but a fading memory and the China Travel Service was beginning to offer guided tours of a few days in Guangzhou (old Canton) which lay 100 km away across the border. Joan and I signed up and joined

149

a mixed group aboard a Kowloon-Canton Railway train at Kowloon Station bound for the Chinese border and beyond. Served with green tea and little cakes by a smart young woman from the mainland in an immaculate uniform, we rattled through the New Territories, across the border and then through almost empty country on the other side of the border. It looked a bit like the New Territories with paddy-fields, market gardens and the occasional factory with its walls daubed with fading slogans; reminders of the frenetic revolutionary passions of only a few years earlier. Our hotel was spotless but grey and grim, the furnishings redolent of pre-war days. Dark red velvet seemed to be the interior designer's favourite fabric, it was used for curtains, bedspreads and cushions, the only colour to relieve the drab walls and woodwork. However, the meals were delicious and helpings were large. We were shown schools and farms, children chanted their lessons and peasants proudly showed off their preserved fruit and their farm-machinery.

Joan in Guilin.

Everyone seemed cheerful and looked healthy, nobody was obese, everyone smoked. On a river trip I fell into conversation with an old seaman who had learned some English aboard ocean-going ships in years gone by. He asked me whether I was a worker or a capitalist; I told him that society wasn't quite like that in the West, adding that in my country it was only the capitalists that actually did any work. He thought this very funny and later on during our little cruise on the Pearl River he introduced me to some of his shipmates and made me say it all again. Remarkably, we were left unsupervised and free to wander round the city and so we took the local buses into town. On one expedition the bus was crowded and a middle-aged woman offered me her seat, again speaking English. Refusing, I told her that in my country no man would sit when a woman had to stand. She replied that in her country a guest is supposed to do what he is told; I sat down while she stood and smiled at me. Europeans had not been seen around much in Guangzhou for 30 years and clearly we were welcome even if only as curiosities or perhaps as a sign that things in China were changing for the better. This was the first of many visits to the PRC. Over the years we came to know many different parts of the country, often joining groups from the Friends of the Art Gallery visiting archaeological sites, museums and ancient buildings all over the country.

CHAPTER 21. GAMMON HOUSE.

Expatriates from all nations, east and west, made up the majority of our patients. We cared for several interesting characters who had arrived immediately after the war and who were now approaching or past pensionable age. Several had arrived aboard ships during the months after hostilities ceased in the Pacific to find a Hong Kong completely messed-up by the defeated Japanese who, uncharacteristically, had not wasted much effort keeping the place clean or maintained. The sewage system and water supplies had broken down, the banks had been emptied of money, the docks were wrecked and the population had been starved, ill-treated and driven away. Fine buildings had been abandoned and desecrated, damaged by the weather or with floors and fittings ripped out. As the Japanese had only imported enough fuel for their own use the locals had stripped abandoned houses for timber and the hillsides for trees to burn for cooking and heating. The invaders had even taken away a large bronze statue of Queen Victoria[37] for scrap, it was eventually found, only slightly battered, in a junk-metal yard in Japan. The Allies had not been too careful either, causing quite a bit of damage when they bombed and strafed during battles against the occupiers.

These men arrived at the beginning of the peace and saw the opportunities and, abandoning plans of continuing to their home countries in Europe, they stayed and worked to rebuild a new Hong Kong, like a phoenix from the ashes. A Swedish friend arrived in 1946 and set up as agent for a Scandinavian manufacturer of cleaning equipment. A German who had become stranded while travelling in the Far East at the very beginning of the War started importing vehicles from the fast-recovering Germany. Others started up factories to manufacture products to send to Europe. These were joined a few years later on the fall of Shanghai by expatriate and Chinese businessmen setting up again after the communists had driven them out from the treaty port. The old joke went; 'He arrived in Hong Kong without two coins to rub together... but, he did have a ship loaded with gold ingots and all the latest textile machinery'. These businesses became the bases of great fortunes, some of their owners

---

[37] Victoria was recovered and she stands today in a public park in Causeway Bay.

eventually living in great spreads along the south coast of the Colony, in Shek O near the golf club and especially on the Peak. These were the old rich, and together with the senior people in the forces and in the government formed a white aristocracy.

There was of course an even larger Chinese aristocracy in Hong Kong, some of it dating back to the birth of the Colony, and many more arrived from China in 1949. Perhaps they were less visible but certainly they were at least as influential in the real business world of the Colony. They wielded political influence through long-existing Chinese groups like the Tung Wah Hospital Board, the various 'Kuks'[38] and I suppose one has to include here the illegal secret societies like the 14K, the Triads which controlled the criminal world.

To start with, Hong Kong, in the mid-1800s was a typical British Colony and was run for the benefit of the colonisers. However, as governor succeeded governor the times and attitudes gradually changed with more and more Chinese representation on government committees and in the completely undemocratic Legislative and Executive Councils reflecting the reality of Hong Kong society as wealth moved into and through the Colony. Of course, while there were few if any truly poor white people in Hong Kong, the really very wealthy were almost exclusively Chinese. These new men and women are typified by Li Ka-shing. He was a high-school drop-out from Canton and as a teenage refugee in 1950 he arrived and set up a plastic-flower factory in Hong Kong. He is now among the 50 richest in the world.

These fabulously wealthy people still live in the background, occasionally to be seen at fund raising dinners and in their very exclusive private clubs. Their young family members may be visible in society magazines and driving round in Italian sports cars but Dad hides himself away and wields his influence in Beijing and at Government House.

---

[38] Chinese friendly societies. Charities doing good works supporting people from particular parts of China, helping abandoned single women or representing owners of farm land.

On the other hand, the poor in Hong Kong, like anywhere else in the world, are in full view in the streets and on public transport. Visitors are sometimes shocked and inevitably see a racial divide, with the rich expatriate trampling the poor Orientals underfoot and forcing them to live in appalling tenements while the white man has a huge flat in Repulse Bay. This is hardly the true case as opportunities are the same for all of us nowadays, although at the time when I arrived in 1974, prior to the enforcement of the government's localisation policy, there was no denying the existence of a bias in favour of expatriates. We were however at the time still very much needed for the skills we brought with us. Today these skills are available in the local population and there are less opportunities for foreign experts. One must also understand that for the expatriate who fails there is no safety net, his family are thousands of miles away, there is no unemployment benefit and he has to leave with his tail between his legs.

Undeniably there is a divide between the races, between the Chinese and the expatriates. I have always seen this as a cultural divide especially involving language; I can never be Chinese, not because of racism but because I do not have the lifelong experience of their culture, their language, their world-view. I hold no animosity towards them even though I am largely excluded from much of their society. Language is indeed a problem; most of the population speaks little English despite almost a century of universal free education for all. On the other hand, Chinese is a difficult language to learn. When visiting most countries, I can pick up a few phrases in the local language immediately, I read the advertisements and signs and make a good guess at their meaning. As hundreds of Chinese ideograms need to be learned individually to garner the sense of a few simple sentences, one cannot just pick up a newspaper or read a notice and start to make sense of the language.

Are the Chinese themselves racist? This is a difficult question that I can only answer by saying that in everyday life in Hong Kong I was and still am well aware that I am routinely discriminated against because I am not Chinese. But is that just lack of communication, that people cannot be bothered with me because it's too hard to relay something that they would need to say? Perhaps a story about Sascha at age 2 or 3 might begin to

154

explain. She has been playing outside with the children from neighbouring flats and I asked her about a new child on the block, Edward. 'I don't like Edward because he's Chinese' she states baldly. I am shocked at this in one so young and ask about another Chinese child, Gilbert, who is one of her great mates. 'Oh, but he's not Chinese' was her answer. For her it was a matter of communication, Edward could not speak English yet while Gilbert was fluent.

Intermarriage between the local Chinese and foreigners is common enough and goes without remark. In the 1970s there was a great excess of expatriate men for the few young European women available. Fifty years previously it was almost a scandal for a white boy to marry an Asian girl but by 1974 nobody gave it a thought. Funny stories abounded of single men with large flats provided by their employers getting married to a local girl and then sharing their home with the extended family of their new in-laws, but I do not recall any of such husbands that I knew being too upset by their situation.

Hong Kong remains a multi-racial and multi-cultural place. The great majority of the population are Han-Chinese and at worst they are happy to simply ignore those non-Chinese who live among them, but on the whole I think that, like me, they feel a leavening of foreign culture, language, ideas and food is an important aspect of the local scene. Naturally they prefer their own but realise life might be rather dull without all the foreigners and their influence. Perhaps the reason that there is so little inter-racial tension is simply statistical; the local population of Han Chinese is so dominant numerically and culturally.[39]

Having mentioned Li Ka Shing I would like to mention a story, probably apocryphal, that was circulated about his generosity to Michael Sandberg, chairman of the Hong Kong Bank when he retired in 1986. A few years earlier the Bank famously and controversially sold its controlling share of the huge trading company Hutchison-Whampoa to Mr Li, even providing the finance as a loan to the buyer. Sandberg explained later that

---

[39] In 2016 96% of the total population were Han Chinese and apart from a tiny percentage their first language was Cantonese. Only 50% of the population were fluent in English.

Hutchisons was failing and needed an entrepreneur of Li's ability to turn it round. The rumour goes that Mr Li arranged a farewell banquet for the departing Sandberg and during the speeches presented him with a big gold replica of the then new Bank building so heavy that it needed several men to carry it into the room. If this was not magnificent enough, Mr Li in his speech added that unfortunately to incorporate the detailing of this model meant it had to be made from 98% purity gold whereas his gratitude to Michael was 100%. Then two lions, replicas of the ones standing outside the bank were brought in; these were of pure gold. Mr Li certainly had a lot to thank Mr Sandberg for but this was munificence indeed. Subsequently he went further and named his scholarship programme after his benefactor and friend.

Doctor Leo Wong, a very fashionable and busy obstetrician, began asking me to attend some of his deliveries and take over the care of the babies as this was now becoming the usual practice at most hospitals. This was very welcome and I came to know a whole different group of families less centred on the Repulse Bay area. Although it is common practice nowadays, Leo's method was controversial at the time. He called it 'the three Bs'; Block, Bottle, Baby. This meant that he induced labour about a week before term and so a delivery date could be booked weeks ahead with everything planned. For the more traditional Chinese families an auspicious birth-date could perhaps be chosen. The mother-to-be arrived in the morning at the hospital and Leo administered an epidural anaesthetic block himself, started an IV bottle of Syntocinon to make the uterus contract and by the evening.... a baby! With his heavy workload the irregular arrival of babies when they were ready would have been impossible. He was an excellent doctor and had little trouble with his methods. A lesser man might not have been so fortunate.

Nowadays when we think of all the health cults, we assume they are a new phenomenon but back in the 1970's we had them too. In an apartment block in Pokfulam where one of the international schools housed their staff, several teachers formed a self-help natural-living group and started to deliver their babies at home themselves. Being young and living a hippy life they conceived several pregnancies and they gradually became more confident about their midwifery skills. I had spoken to one pregnant teacher about this, implying that inevitably, eventually, something would go wrong even with the 'natural process' of childbirth. One day it did and they frantically called our clinic, I arrived at the flat where the mother lay bleeding with retained afterbirth; they had refused to call an ambulance and she was very ill and as white as a sheet. Fortunately, I was in luck and a firm pull on the umbilical cord to release the retained placenta plus a shot of ergometrine stopped the rush of blood. It could have gone otherwise and then the new baby would never have known its mother. Naturally this failed to convince them of their folly and, probably because I made clear my anger at their foolhardiness, they never called me again.

Another group of single-minded people was a breast-feeding group. These women, and a few men, were committed opponents of bottle feeding. Their attitude to lactation was that it was almost a holy rite rather than a rather pleasant and convenient way of feeding a baby. I agreed to enter their lions'-den on one occasion to give a brief talk on baby feeding and recent scientific research on the matter. The audience, many busy feeding their youngsters viewed me cynically. One child of school age kept running in from playing outside every few minutes to lift his mother's brown coarse-knit jersey to take a quick suck at her enormous bosom, then ran outside again. This was not my scene and they did not invite me back.

One of my patients was a Christian Scientist; although she brought her children to me for conventional medical care[40], she and her husband followed their faith and when she developed ovarian cancer, she refused medical treatment. She preferred to go to her death surrounded by her fellow faithful who read the scriptures to her for hour after hour. I visited her a few times on her death-bed but she even refused morphine for the dreadful bone pain. I still worry that my offers of pain relief may have been interpreted in her suffering as temptations from the Devil and caused even more distress than if I had kept quiet.

We were growing busier, and, needing more space, I had persuaded Bill that we needed to move from the cramped rather gloomy rooms in the

---

[40] In this she was being sensible. Parents had been prosecuted and their families taken into care for refusing conventional care for their children.

Chinese Bank Building to a larger and smarter new office in Gammon House[41]. This gave us room for a better waiting area and two consulting rooms as well as a small lab with an X-ray machine. The new office attracted many new patients including businessmen from nearby office blocks; also, the nearby carpark was convenient for mothers with young children. Vivienne managed the new clinic and we employed an English trained nurse and also a radiographer, Winnie, who doubled as dispenser as well; our little practice was growing.

Several insurance companies now engaged us to perform check-ups on their clients and a couple of consulates asked us to verify the health of potential immigrants. This work was repetitive and rather boring; neither Bill nor I liked doing it very much as it was quite exacting as well. Vivienne, as practical as usual, ticked me off one day when I was moaning about this extra work-load, pointing out that per minute spent doing such examinations we were earning several times our regular fees. Applicants for life insurance had to be warned that I had no duty of patient confidence since I was being paid by the insurance company to extract information. I would point out that they had to answer my questions honestly but concisely otherwise adverse information might be revealed that I would have to mention to the company. The questions on the forms were searching but did not cover everything so it was important that a garrulous client should not admit to anything not specifically enquired about or he might face refusal or higher premiums.

One of my early clients (for they could really not be regarded as patients) in the clinic for an insurance medical, a young Chinese man, vouchsafed to me that this was his third examination before his life policy would be approved. I expressed surprise and he added that this was because it was for a high sum, 5 million US dollars, a lot of money in those days. He added that he was getting married and his bride had insisted on this insurance. I rather thought that I personally would be quite worried sleeping all alone with this lady with such a price on my head.

Generally, medical examinations for immigration purposes were straightforward. A chest X-ray, a physical and a few questions and then I

---

[41] Later renamed Bank of America Tower.

could complete the official forms and sign at the bottom. The X-ray was sometimes problematic. One day I had to tell a young man that his film showed active tuberculosis and he should go to his own doctor straight away. Unfazed he said not to worry because in fact the X-ray was of his brother who had stood in for the test so could we not just repeat it? I felt his chosen country could well do without a new citizen this devious and reported what had happened to the consulate; they took him anyway.

Another well-paid chore was doing check-ups on merchant-seamen for one of the shipping lines. It was stipulated that any sailor with a stomach- or duodenal-ulcer could not continue to work at sea. Since heavy drinking was common amongst these men, we quite often diagnosed such ulcers. Then a decision had to be made whether the sailor should have an operation or should adopt another profession. Usually, they chose to have surgery. Nowadays of course treatment with over-the-counter tablets effects a complete cure and these major operations and their frequent long-term side effects are consigned to history.

Then came the Vietnamese Boat People. They started arriving in small numbers in Hong Kong waters aboard rickety old ships to escape the new Communist regime ruling their country. The British High Commission asked Hong Kong's private doctors to help by examining these unfortunates to help speed their processing as refugees. Bill and I arranged to see 30 or so of them daily and opened our lovely new clinic for a few hours in the evenings for the purpose. Winnie, Vivienne and the nurses were delighted to be paid overtime and soon the Vietnamese would arrive in a bus to be examined and X-rayed. From the start it was chaos, nobody at all seemed to understand waiting in line and they all absolutely refused to even undo their clothing to be examined. It did not take long to understand their reluctance; I had just weighed a little old lady and could not believe the scales. When I had wrestled her clothing aside to sound her chest, I found she was festooned with rolls and rolls of gold-leaf sewn into tubes of silk material. The family's entire wealth in gold was wrapped round grandma; obviously their hope and their future investment for a new start in the west. We had still more arguments when they had to shed their weighty treasure to be X-rayed. The babies were also a problem; apparently in Vietnam babies do not wear nappies but are carried with their bottoms exposed to the air, a small cloth held strategically. This was obviously good for the babies' skin but not so good for our new wall-to-wall carpeting as their mothers would simply hold the baby out at arm's length when he wanted to relieve himself. Vivienne quickly solved this by packing a disposable diaper round anyone under the age of five.

After a few months the government and the UNHCR took full responsibility for the boat-people as they arrived in increasing numbers often after terrible voyages north from their home country. Attacked and robbed by fisherman on the way, there were murders and rapes and the people smugglers, 'snake-heads' as they were called, also mistreated and swindled them. This human disaster continued for many years but was hidden from view in the 'closed camps' where conditions were gradually worsened to encourage the refugees to accept what amounted to a bribe plus an air ticket back to Saigon, now renamed Ho Chi Minh City after the almost holy man whose philosophy had rescued his people from the evils of French, then American imperialism.

One little boy seemed to come to the clinic quite often; his parents had paid for medical insurance for the boy and they seemed to be getting their money's worth as he was in and out of the clinic with various minor ailments all the time. One afternoon he arrived with a sprained wrist after a fall. After sorting this out I decided to look at his ears to check the progress of the ear infection for which he had visited me a few days earlier. As I picked up the otoscope his mother became upset and suggested I did not really need to do this, she then blurted out that she in fact had two children, identical twins, and she had thought that if she only insured one of them how would anyone know? Now she thought I had rumbled her stratagem, although I probably would not have done so but for her confession. As one might expect she could not face me again and she did not bring her boys to the practice any more and presumably found another doctor to trick.

I became a member of the Round Table, a young men's dining club with branches all over the world. It was a good way to make friends from different walks of life and most of its members were from the legal and surveying professions with a few senior policemen as well; we raised money for various charities and did various good-works like painting an old-folks-home and taking deprived kids on beach picnics. Through Round Table I was asked to take on some pro-bono medical work for a German charity that provided services for the blind including a school and also a home for old ladies. The latter was run by Lutheran nuns. Chan Yee was the oldest inmate there. She had been blinded by smallpox in her teens and was in her nineties. When new patients to our clinics asked what was my upper age limit for patients, since often paediatricians refused patients above a certain age after puberty, I would point out that my oldest patient currently was 96.

Several of the ladies had the tiny deformed feet that resulted from binding. This cruel practice had finally been banned in China in the 1920s under Sun Yat-sen's government; but here, fifty years on, there were these old women who must have been from once well-to-do families in the last years before the fall of the Ching Dynasty. In one way they were living relics, but in reality they were lively old people who were enjoying their waning years in this little corner of modern Hong Kong and despite their blindness seemed determined to have a good time.

My visits to the home were busy ones as the residents liked the chance for a chat with the young doctor. At Christmas they knitted the most beautiful jerseys in intricate patterns for our daughters. On one memorable evening they invited Joan and I out for a Chinese dinner at a restaurant in Wah Fu, a nearby low-cost-housing-estate. Some of the dishes they chose consisted of the more interesting parts of various animals which are so much loved by Chinese diners. In the west these organs would end up in pet-food but in China are reserved for the guest of honour. It was a relief that my hosts could not see me pushing such morsels onto the edge of my plate.

The blind school's head-master asked me to see one of their pupils. He was an eight-year-old boy who, as a result of meningitis five years before, was not only blind but had epileptic fits. After his anticonvulsant medicine had been changed a year or so earlier, he had turned from a nice little boy to an absolute terror. Permanently angry and disobedient, he was often violent and had become unmanageable. This is a recognised side-effect of

Phenobarbitone, the medication which he was taking. I advised a change in prescription and wrote him up for Valproate. Almost overnight he changed back into the sunny little chap that he had been previously. I saw him a few more times to get the dosage right and signed him off to return to the government clinic. I thought no more of him until I received a visit from his father, a flashily-dressed well-built man. He thanked me for saving his son from inevitable expulsion from the school and said he would like to reward me. He showed me a thick envelope stuffed with currency, I told him that I could not accept it as my work at the school was a charitable contribution, so if he wanted to give anyone some money, he should donate it to the nuns for their work. 'In that case' he continued 'perhaps I can help you. Do you have any enemies?' At that moment I realised he was a Triad gangster and imagined him breaking the legs of a few people who had given me difficulties. With some regret I refused this offer too. As he was leaving, he said that he would phone me from time to time with useful advice and thus we parted cordially. A few weeks later he rang, 'Number 3 in the fourth' he said and rang off. Ah! A horse racing tip. After the race I checked the result of the fourth and number 3 was third. I did nothing with these regular tips over the next few months until he called again to ask whether I was happy with his advice. No, I wasn't as they never won. He then patiently told me that I was supposed to make a combined 'quinella' bet with both the favourite and with his horse. He clearly regarded me as a complete innocent. The next time he called with a tip I did as he told me. I asked Bill, who was a Jockey Club steward and would be at the racecourse on the day, to place the bet for me. On the Monday after the race he was furious. 'Don't ever ask me to do that again' he fumed. He had been very embarrassed when my horse after running last for most of the race was allowed through to finish third as the rest of the field seemed to slow down in the last furlongs. The stewards' enquiry was undecided but let the result stand, after which Bill had to go and collect my very suspect winnings. That was the last time I heard from my Triad friend.

At the time a popular way of thanking a doctor was to give him a voucher to buy a cake. Marias was a chain of cake shops that sold such vouchers and inevitably I started to accumulate a sheaf of them in my desk drawer. Joan and I were not fond of the type of cakes that Marias sold and so I took to giving them to the blind ladies which added to my popularity in that sector. At around this time Hong Kong had a 'run' on Maria's cake shop coupons. It started when a newspaper article claimed that the value of unused vouchers in circulation now exceeded Marias' total asset value and they were about to be dishonoured. This caused a general panic and the cake shops were besieged by mobs of customers brandishing their vouchers and demanding their cakes. In the end Marias survived and I imagine everyone had their cakes and ate them, but it was illuminating to see how a tiny rumour managed to create uproar and panic for a week or so. Other places have had runs on their banks but I'm sure only Hong Kong has ever had a run on cake-shops

I enjoyed the small amount of charity work I did and through it I encountered a different world with different and perhaps simpler values. There was a gardener from the Golf Club. I cannot remember how he first came to our office but he would turn up from time to time over a period of twenty years. I couldn't possibly charge him our regular fee, but on the other hand he would be insulted if he felt he had been given charity, so I had to give him a bill at a fraction of our normal fees. Vivienne took to calling such patients my 'Lame Ducks' and drew a cartoon duck with a walking stick on the front of their notes to remind the staff that they shouldn't be billed by mistake.

Another 'lame duck' was the divorced wife of a local airline magnate. A very pretty Chinese woman. When they were still married, I had helped them adopt a child of about 11 months, I had seen the little girl at the orphanage and certified that she was normal and healthy despite not then being able to sit properly nor crawl yet. I explained to them that her apparent slow development was due to enforced immobility in the orphanage, where children were restricted for months on end in their cots at an age when at home they would be cruising round the house. Indeed, within a week of moving into their flat she was tearing round the place and keeping everyone in their place. After the divorce mother and child lived in rather poor circumstances and eventually moved to Istanbul with a young Turk she had met.

Adoption at the time was a rather informal affair and followed Chinese family customs where perhaps a couple with an excess of children would

163

hand one or more of them over to other less blessed family members to raise. The law would then regularise the arrangement once it had lasted more than a year and if everyone agreed. Later, adoption became formalised with social workers and officialdom involved from the start.

I was consulted by infertile couples from time to time and would start some investigations prior to referring them to Bill for definitive management and treatment. One night the senior nurse at one of the hospitals rang me with an unusual question, 'Did I know any patients who wished to adopt a baby?' It turned out that there was a young woman in the charity ward in labour who did not want to keep the child. I had to be quick and the baby had to be removed from the hospital that very night. I rushed to the clinic and found the notes of an infertile couple I had recently seen; Helga and Max, from Germany. They were childless after several years together. I rang them at 2 am. Surprised, shocked even, they quickly understood and said yes.

Bring warm blankets and some cash, I told them and we met in the carpark outside the hospital and were ushered into a private room. The baby girl was perfect and very wide awake, her limpid black eyes looking around at us all. Helga lost her heart on the spot. They handed over the money to cover the mother's bill and drove home armed with a few disposable diapers and some sample bottles of milk. A year later that baby became a German citizen and she remained my patient for several years until they returned to Europe.

I was involved in several adoptions like this, though they were usually better planned in advance. I came to know a Christian lady who used the alias of 'Mrs White' to hide her true identity. She was an associate of the famous missionary Jackie Pullinger who worked assiduously among the poor and the drug addicted in the dreadful Walled City in Kowloon. This slum area of tenements was theoretically still administered by a magistrate in Canton. By a strange anomaly the law in the Walled City was still Chinese and the Hong Kong government had no jurisdiction there. It had become a robbers' den of crime, prostitution, drug dealing and, for some reason, unregistered dentists. I entered under Mrs White's protection. We dodged open sewers, festoons of live power lines and the hostile stares of the rather daunting denizens from doorways and windows. With no building regulations the tall unkempt buildings were separated by footpaths only a metre or less wide. Water lay in stagnant puddles and live electric wires drooped from poles. Mrs White's mission was to rescue unwanted babies and have them adopted away from this horrible place. I would go into flats with her and examine the children she had identified. Unfortunately, some were not well and showed signs of congenital damage from their mother's drug or alcohol addiction, while others were malnourished or had infections, but most were fine children and we were able to help them. Boys were in greater demand of course. The natural mother expected to be paid and though this was illegal it was always somehow managed. The big problem was that the natural mother had to finally sign away her rights when the child was a year old. If this was mishandled the adoptive parents could face blackmail at this point with a huge sum of money demanded. It was the reason that the identity of all parties had to be kept secret while Mrs White acted as negotiator; a function she performed expertly. Once the child was safely away from the awful slum city and living in its new home the adopting parents would eventually have to go before a magistrate who after instigating some inquiries could order a new birth certificate to be issued.

Nowadays there is none of this cloak and dagger excitement, it's just a bureaucratic paper trail and so much the better, but it is also much less fun.

The Walled City is no more; it was demolished in 1990 and on its site there is now a public park laid out like a traditional Ching Dynasty formal garden.

CHAPTER 22. THE PRACTICE GROWS.

In the 1970s the English language TV programmes aired in Hong Kong, were, with a few exceptions, dreadful. The upside was that we wasted hardly any time in front of the box. The Mercedes-Benz dealership however, on Sunday evenings, sponsored high quality dramas and serials, usually made by the BBC. They were not interrupted by adverts, not even for cars. Generally, TV ads, mainly for Cognacs or fancy watches, were quite awful and widely mocked[42]. One favourite was for air-conditioners in which the local agent, a prominent member of the Yacht Club, looked the camera in the eye and told us that, as he was an 'engineer,' we should trust him and his products. On the other hand, the three English language radio channels were very popular with a choice of light entertainment, classical music or the BBC World Service.

While driving to work at breakfast time the expatriate community listened to Ralph Pixton's Open Line phone-in show. Pixton was a loveable rogue; perhaps more rogue and less loveable, as we would learn much later after his death in 2001. It was said that he owed large sums of money that he had extracted somehow from friends, who often could ill afford it, and which he had never repaid. He had arrived in Hong Kong as part of an itinerant repertory theatre in the mid-1940s and he never left, making a living from his actor's trained voice. A notorious TV advert he voiced over in the 70's for Athlete's Foot powder was still being aired over thirty years later.

Open Line had listeners calling up about holes in the road, people illegally chopping down trees or dumping rubbish, barking dogs, stray cats and everything else. There were often enquiries about the wild animals which abound even in the urban areas; snakes, muntjacs, porcupines, civets and more. One caller from the New Territories complained of a lizard in her sitting room, Ralph told her it was called a gecko and that they often ran round the walls and ceilings in your house eating insects. Adding cheerily that they were charming little things and quite harmless. A bit later she rang again and said she was still worried as it had now knocked over

---

[42] See https://www.youtube.com/watch?v=YShUCXs8Gjl

her coffee table. Pixton urged her to call her district-officer to remove the iguana, or was it a monitor lizard?

He had a running joke going; whenever Mr Romer, the government's wild-life expert appeared, Pixton played a tape of twittering birds in the background. Romer was a serious biologist and even had a species of small frog named after him. He was a great snake enthusiast and frequently turned up with the government's animal catchers to help residents extract pythons from drains or drive cobras out of kitchens.

When Pixton was on leave various announcers stood in for him, but they were pale imitations of the master. One of them was having an affair with an acquaintance of ours who had a very distinctive Irish brogue that we all recognised when she rang the programme to tease him, revealing that he liked her to dress in a girls-school uniform on their assignations and slipping in other salacious details.

For most of the day the channel had played dreary pop music and so I kept to the World Service. The bilingual classical music channel was so highbrow that it defeated me. Shostakovich quartets and Teutonic ladies singing lieder were just a bit too much before breakfast.

I had to spend a lot of time driving around Hong Kong Island, not only to the hospitals and between clinics but also on home visits. These are not done much, if at all, nowadays but in the 1970s and for the next twenty years we gave out our home phone-numbers to our patients so they could reach us after hours. Indeed, visits are time consuming and there is a limit to what a doctor can stuff into his bag, so that a necessary drug or bit of equipment could be lacking and so prevent him from doing the very best job. But for someone immobilised by sciatica or a vomiting child, for example, visits were very useful. I recall two cases of meningitis that I was able to diagnose at an early stage and send to hospital who may have otherwise had to wait for the clinic to open had I not gone to see them at home. Stitching up minor cuts was something else I would do on a patient's kitchen table. I carried a sterile pack of instruments in the car and usually had the wound stitched before the natural numbness around the injury had subsided, the risk of hospital cross-infection was obviated too.

However one had to have a sensible parent to restrain the victim in the absence of a nurse to assist.

Sometimes home visits were not such a good idea. One afternoon I was at the Canossa Hospital when a lady patient in an overlooking block of flats saw me in the car park. She rang the hospital reception desk with a request for a visit when I had finished there. I walked over to her block and went up in the lift. She greeted me at her front door wearing very little and what she did have on was totally transparent. She looked extremely hale and hearty, so, being a bit of a coward, I beat a strategic retreat. I could foresee a lot of trouble had I stepped in there without a chaperone. Chaperones after all are mainly for the doctor's protection.

Unless he has the single-minded dedication of Dr Wong, (my obstetric colleague celebrated earlier for his 3 Bs approach to delivering babies), there is a limit to the amount of maternity work a lone doctor can undertake, and even my indefatigable partner Bill was feeling overworked. So, when John Elias-Jones sailed into our lives I worked hard to persuade him join our practice. JEJ was a very Scottish gynaecologist spending a year of his training at Queen Mary Hospital. I had met him at the Yacht Club where he often crewed on one of the 505s. Strongly built, tall and an able sailor he was the perfect front-man in a trapeze dinghy. His true nautical metier was on the foredeck of a big racing yacht and he was among the crew of *La Pantera*, then the local champion offshore racer.

Professor Ho Kei-Ma was the redoubtable doyenne of Os & Gs at the time and JEJ had just completed a stint under her tutelage. He wanted to stay a bit longer in the Colony despite his need to return to Scotland to complete the last year of training for the essential MRCOG qualification. He moved in with his 505 skipper Hugh Blaik and his wife, and joined the practice where he was an immediate success. He had enormous charm and sympathy and women, whether pregnant or not flocked to him. Under Ho Kei-Ma's guidance he had become very skilled in cancer surgery, an area that was developing rapidly at the time. Bill asked him to take over some of his more challenging patients and John's reputation soon spread among the local medical community. Pregnant women were soon coming to him for care and he was able to stand in for Bill sometimes, allowing us all a bit more leisure time.

In his short time with us he became a close and loyal friend but common sense prevailed and after two years he returned to Glasgow leaving a big gap in our lives. We remained firm friends and he often visited Hong Kong as visiting lecturer and examiner at the Medical School. I sailed with him on his yacht *Stardust* in the Western Isles of Scotland several times over the years before his untimely death from cancer.

John Elias-Jones with Sascha and Julia.

JEJ was a great bon-viveur and a raconteur. He was often at our dinner table in Shouson Hill talking and drinking malt whisky into the small hours. Then it was too late for him to go home so he would sleep in our spare-room and in the morning borrowed a crisply ironed shirt from my wardrobe to set off to work. In his last week in Hong Kong he returned them all in a creased bundle. Such are bachelors.

Time passed, and in 1978 our third daughter Julia was born again with Bill in attendance. He had been a bit miffed with Joan and I on the previous weekend when we had gone out sailing on the 505. Joan was leaning right out over the water using the trapeze, the abdominal swelling that was to become Julia no doubt adding to our stability. By chance Bill was aboard a passing motor-boat and spotted us tearing along in a welter of spray. In his rather old-fashioned view ladies in advanced pregnancy should stay at home.

With a growing family we decided to move to a larger flat in Ann Gardens, just down the road, but still in Shouson Hill. It was much smarter and had a better layout. The private road outside was away from the traffic and Sascha and Diane could safely ride their bikes and play with the other children living nearby. It suited Spitz too and moved him further from the police driving-school where he was in their bad books. Spitz hated both motorbikes and their riders. When in their wisdom the police sited their motorcycling school near our home, they reckoned without our little fluffy white dog who spent a lot of his days sitting under a hedge beside the road waiting for them. Led by an impressive sergeant on a weighty machine the learners wobbled along behind on low-powered bikes in a long double file. Spitz would bide his time. At the critical moment, he emerged barking, going straight for the foremost learner who would swerve and tumble off. The others tried to avoid him but on a good day Spitz had the pleasure of seeing a struggling heap of policemen lying in the road under their capsized bicycles. The sergeant would stand there, notebook in hand, wagging his pencil at the dog. Spitz was never arrested but I believed that it would not be long before the police came after his owners.

On the evening we moved into Anne Gardens our downstairs neighbour came to visit us. He knocked on our door, and we thought it was a neighbourly gesture to come round and welcome us, but actually he had come to complain of the noise as we moved furniture and unpacked boxes. He did not soften in all the years we lived above him. He and his family we dubbed 'the Polishers' as it seemed that everything they owned was white and gleaming; cars, children, water-skiing boat, hair, teeth, everything, even their Alsatian bitch. We hated them and the feeling was mutual. The only time they showed any friendly interest in us was when I borrowed Bill's car for a week. The gold Rolls Royce looked very impressive in our car-port while the Austin was away for repairs and Mrs. Polisher became quite chatty. Then the Austin came back and we were cut off again.

The local beer was delicious and cheap, especially after Carlsberg had set up a big new brewery in the New Territories in competition with 'San Mig' whose slogan was 'A real friend', to which one could add 'except for the hangovers'. Wine however was quite expensive and not yet a popular drink among the Chinese community. The better-off locals liked an expensive Cognac, often diluted with lemonade. The ordinary man in the street had Mau-tei, a rice-based liquor with the kick of a horse and the bouquet of diesel fuel. In the entrances of the better Chinese restaurants were cabinets displaying Mr Wong's bottle of Courvoisier or Mr Chu's half empty bottle of Remy Martin and many more; some of them worth several thousand dollars and all waiting for their owners to come in and dine.

Joan and I liked wine but its price made it an occasional luxury until we joined a group of fellow impecunious wine-lovers who under the label *'Chateau Bon Vivant'* bottled their own. Twice a year we subscribed to a couple of steel barrels from France, like oil drums, which were delivered to a friend's car-port at Plantation Road on the Peak. We took our own washed and sterilised bottles and filled them using a siphon. We had a corking machine and labels emblazoned with the *Chateau Bon Vivant* name to stick on the bottles. Usually we chose a red and a white from Bordeaux or Burgundy; they were significantly better than the supermarket plonks like *Hirondelle, Blue Nun* and the Chinese vintage *Dynasty* which everyone dubbed 'De Nasty Wine'. Finding 36 empty wine bottles was a challenge that I solved by raiding the bins behind the kitchen at the Repulse Bay Hotel one evening

Hong Kong has its own symphony orchestra which gives very professional performances at City Hall and the Cultural Centre. Famous conductors appear and we can hear soloists from all over the world, including from China. A full-blown opera is staged annually and the musical arts have a powerful and knowledgeable following among Hong Kongers. It was not always so. When we arrived in the 1970s the 'Phil' was in its infancy, although local players did their best under the batons of part-time maestros. Anyway, that was all we had if we wished to hear live performances and we loyally turned up twice a month for an evening of missed cues, varying tempos and what the local critic called the strings' 'Mantovani effects'. Listening to 33rpm LPs on the gramophone at home was one thing but even with our performers, hearing the same music from a full-sized orchestra was a great thrill. One autumn Georg Solti brought the Chicago Symphony on tour to Hong Kong and we attended one of their performances. The tiny Sir Georg conducted manically and the orchestra played wonderfully, it was an expensive night out and we disloyally felt that the 'Phil' did sound a bit flat for a while after that.

Spitz too was a great music fan, he clearly preferred sacred music and whenever I played the *Rex Tremenda* from the Berlioz Requiem he would howl along with the massed choirs. Even today when I play this music I hear in my mind the howl of that long dead dog.

For adult expats there was a lively cultural scene with local and visiting performers. The Hilton Theatre which put on performances with such luminaries as Brian Rix and David Nimmo. Dave Allan and Dame Edna Everidge (Barry Humphries) appeared at the Mandarin dinner-cabarets and of course we had our watering holes in our clubs or favourite hotels. But for teenagers and young adults there was very little provided specially to suit their tastes. Even the Discos were expensive and unwelcoming to youngsters. The result of this was that when they became bored with trailing after their parents or playing sports, all there was to do was to wander round the streets or hang-out on Repulse Bay Beach near the 7-11 store where they could buy beer and cigarettes if they could persuade the shop assistant to believe they were over 18. Getting into 'bad company' was inevitable, perhaps leading to drug abuse and under-age drinking.

At Round Table we discussed this problem and one of the members, Harrison by name, who had somehow inherited some equipment offered to run a disco once a week for teenagers if we could find a venue. After some research, we settled on the Seven Seas Bar, a girlie bar in the red-light district of Wanchai. It had a large dance-floor and the mama-san offered us the place from 8pm to 11pm every Friday night. In those three hours the bar would only serve sodas and the 'girls' would stay out of sight. It was perfect and *Wings-Disco* was born. When it was our turn to supervise, Joan and I would head for Wanchai and stand at the door of the bar taking entry fees from the youngsters who dashed inside to the roaring music and flashing lights where they hopped around merrily with their pals, fuelled by no more than Pepsis and Cream Sodas. I would scout round the floor checking for hip-flasks and drugs. Even cigarettes were forbidden. Younger children were collected by parents while the older ones wandered off at 11pm. It was a huge success with the kids, and the bar was pleased with their takings. We raised a fair sum of money for our charities too. I came to know the mama-san over the time that Wings Disco ran and for many years after she would greet me in the street by name with a familiarity that impressed visitors no end when I accompanied them to the Wanchai bar area which possibly gave me an unwarranted reputation as a dark horse. Like all good things *Wings* closed down after a year or so as Harrison wanted to move on to other ventures.

Benjamin Franklin famously said that 'Nothing is certain but death and taxes' and even though income tax rates are low in Hong Kong we still had to pay up. In the UK the government deducts what it estimates you should be paying directly from your pay-cheque every month and I had been used to that. Now in the financial hub where I was living, I was having to remember to put aside 15% of my earnings monthly and as the year went on a significant deposit accrued in the bank. Everyone had to do this. There were many companies, large and small, who wanted to exploit these temporary savings although every April these deposits would revert to zero as the taxman sent out his bills. It's hard to believe in today's climate of low interest rates that at that time even the meanest bank paid 4% or more on deposits like these; so the various schemes these 'investment' companies offered had to have a good line of patter and a better return than HSBC[43], but quite often one heard of friends whose invested money had evaporated from such enterprises.

A long-standing patient, Barry, took me out for lunch to introduce me to a scheme he had invested in. Barry was a well-established insurance broker and had joined this scheme which he said was being offered only to a small group of prominent business people. Although not an expert in

---

[43] HSBC. Hong Kong and Shanghai Bank.

investment or finance, Barry had already persuaded many of his personal friends to put their money in. He explained that the scheme bought and sold on the currency exchanges and, amazingly their methods would make money both when currencies moved up or moved down in the market. A lavish brochure demonstrated this with graphs and pie-charts. I took the booklet home but could not make head nor tail of it. A banker friend had always advised me never to put money into anything I did not immediately understand and I am glad I did not. Barry in his enthusiasm, continued to press me; especially as his deposit had already grown almost exponentially in value and he had now sold some of his portfolio of other investments to add to his contributions in the scheme.

Then Barry disappeared. We later heard that he had lost everything. His family had to sell a much-loved and valuable painting which was believed to be by Chinnery[44] to be able to leave Hong Kong. And leave they had to, since all Barry's friends were chasing him after they too had lost large sums on his advice. I saw this sort of thing several times as greed overcame common sense. In Hong Kong it is not hard to become moderately wealthy, but for many this is not enough. Thus, relatives and friends can be drawn into financial disaster.

A caricature that springs to mind when thinking of the British Empire is of the expatriate drinking a chota-peg or taking tiffin in his Club served by native bearers. Hong Kong too has its fair share of clubs; indeed Joan and I had joined the Yacht Club soon after arriving in the Colony by which time the racial segregation in clubs had disappeared. While Chinese members were still outnumbered by expats they were accepted and indeed serving on committees and fully participating in all the club activities. Some clubs with long waiting lists were nearly impossible for newcomers to join so I was very happy when I heard that the Hong Kong Cub in Central had decided to admit some younger men to their membership; but only men; racism might be in retreat but good old sex-discrimination was still in force! Thus, I and a few of my contemporaries in our thirties were admitted to the hallowed and much-loved old building standing like Miss Havisham's wedding cake in the middle of Central.

---

[44] George Chinnery. 19th century portraitist in Macau prior to the Opium Wars.

Like Miss Havisham's wedding cake, the Hong Kong Club in 1981.

A few years later our family were also admitted to the Country Club near our flat in Shouson Hill and could use its swimming pool and tennis courts. These clubs provided little oases of calm and quiet in the busy and noisy environment of Hong Kong. A place to relax, to meet friends, to take the children; though not to the HK Club where even 'ladies' were restricted in their movements to one dining room and their own sitting room. Later, after the old club was demolished and replaced by a new building on the site, there was great consternation when our wives were admitted to the bowling-alley bar; the sky didn't fall in.

Other nationalities enjoyed their clubs too; the Portuguese had the enormous Club Lusitano in Central, in Kowloon Tong were clubs for the Indian community and sports clubs abounded; football, cricket, my own Yacht Club, the very exclusive golf clubs and the Jockey Club. The Victoria Recreation Club, perhaps the earliest sports club in Hong Kong stood by the beach in Deep Water Bay. The Ladies had the Ladies Recreation Club on Old Peak Road with its rigid rule enforcing regime, and the so-called 'virgins' retreat' at the Eleanor May on Garden Road. All had their different memberships who found within their portals peace and perhaps a feeling of being at home, wherever that might be. I was not invited to any of the Chinese clubs but Tony Daroza often took me to Club Lusitano and its aircraft-hangar of a dining room where we ate African Chicken or dried salt cod.

CHAPTER 23. CHRIS PUGH MRCOG.

When we realised JEJ could stay no longer Bill and I started to look for his replacement; the new doctor had to be fully qualified and hold the MRCOG diploma. At the time one didn't even need to hold a membership or fellowship diploma to practice a specialty. This enabled doctors who hadn't quite made the grade in their home country to set themselves up in Hong Kong where the regulations were pretty easy-going in the true laissez-faire tradition. Such newcomers in the medical and other professions were sometimes referred to as 'FILTH[45]'.

Because of such widely varying standards among the local specialists, there was now serious talk in medical circles of tightening up the rules. Hospitals were becoming more concerned about vetting surgeons and other specialists before allowing them privileges. Since there was no medical authority legally charged with defining and registering specialists, there was pressure on the government to look into this. It was actually against the Medical Council's Code of Practice regulations to call oneself, properly qualified or not, a specialist at all. Business cards and clinic signs were not allowed to mention a doctor's area of interest. Out of this confusion emerged the Academy of Medicine and its various specialist colleges. Before long we doctors were asked to submit our academic qualifications and CV to the specialist college of our choice; I was not certain myself whether to apply to the General Practice college or to the Paediatricians. My dilemma was solved when the Prof of Paediatrics at Queen Mary Hospital told me to get a move on or I might miss out. So, I was accepted as a specialist paediatrician and, after going through the pomp of a formal robing ceremony in the Academy's lovely new building, was allowed to add the letters FHKAM (Paed) after my name.

These new rules meant that our new obstetrician had to have completed his training and had to hold the Royal College's diploma otherwise with legislation on the horizon he might lose his rights to practice his chosen specialty. We advertised in several countries whose qualifications were recognised in Hong Kong, this really meant the

---

[45] Failed in London, Try Hongkong.

countries of the old Empire. We had many enquiries which whittled themselves down as applicants got cold feet, were countermanded by their spouses or in one case just needed to redecorate his house in Stockport. In the end, just as we were thinking our search had drawn a blank, the perfect man applied; Chris Pugh. He was finishing a two year's contract working in Saudi Arabia and he held the all-important MRCOG diploma. He had recently married a Scottish theatre sister and was looking for a long-term appointment, and having been brought up in East Africa did not have strong ties to the UK. He came out to see us and liked what he saw.

Chris had a natural sympathy with women and his energy and open manner soon attracted a good following. A friend observed that all the most beautiful western women in the Colony came to our practice, and looking around I thought that might be true.

Myself, Joan and Chris Pugh.

While searching for somewhere to live, he was our guest at Anne Gardens and took over Julia's bedroom while she shared with her older sisters. Julia took a great liking to Chris whom she always addressed as 'Doctor'. She would call to him every morning at first light, poor man, announcing 'Doctor, I'm wet'.

A message he would pass on to Joan and I. No wonder he quickly found accommodation at the Matilda Hospital's new block of flats for medical staff and moved out. One evening we managed to lock ourselves out of the apartment; but we could see an upstairs window ajar, and Chris jumped into action and practically ran up the wall and in through the window. This feat of athleticism silenced my doubts about the hair-raising tales he liked to recount of his earlier experiences as a Territorial Army soldier in the elite SAS.

Our new clinic below Susi Q, near Clearwater Bay

With two obstetricians working so hard it made sense to employ one more doctor, preferably with general practice training. With three males our set-up seemed a bit unbalanced so a lady doctor was sought. We were fortunate to find Fiona who had recently arrived with her husband from England. He was an engineer with the MTR Corporation at the time building extensions to their existing railway network. Fiona was trained in Scotland and they had two little boys. Our plan was to open a clinic near Clearwater Bay in the New Territories where the rapidly expanding Cathay Pacific airline was housing their employees. Cathay were very supportive and encouraged us to find premises in that area which at the time was somewhat under-doctored.

The new clinic was near the shoreline in Junk Bay where we rented a ground-floor shop next to a DIY store owned by an airline pilot. Nearby was a shipbreaking yard where ships of all sizes were being reduced to scrap and it was always interesting to go there and watch the workers with oxy-acetylene torches cut out huge sections of steel plating which were craned away to be turned into reinforcement rods for building Hong Kong's new housing blocks. There was a showroom too with ships' clocks, brass binnacles and bells engraved with ships' names. Bill bought a great big binnacle with a compass viewed through an oval window and it came to stand incongruously in the entrance to his flat.

We engaged an Australian nurse who was extremely competent but had very fixed attitudes that could drive Fiona to distraction. This lady loyally worked for us for several years. She knew everyone's business and kept the patients in order even when they didn't need it. To put it another way, she was very bossy. Everyone relaxed when she returned to Australia and was replaced by Beryl who was soon joined by a second Beryl. This team of three women carried the Clearwater Bay operation on their shoulders for some years. The Junk Bay shoreline is no longer there as the area was reclaimed and on it stands the huge new towns of Hang Hau and Tseung Kwan O, housing a million or more people and the shipyards and steelworks are now part of history. These changes happened after our clinic had moved a few miles away to a new shopping centre near the prison at Razor Hill. Sadly, Fiona left us when her husband returned to work in London for British Railways. The two Beryls, who had now firmly become part of the practice stayed with us until after I left.

With two successful obstetricians in the practice as well as several independent doctors asking me to see babies, I was rushed off my feet. I was unable to extend my working hours as I was not welcome to disturb the mothers' sleep by arriving any earlier in the mornings at any of the three different baby units where I worked.

With Beryl and Winnie.

Caesarean sections were generally scheduled for 7.30 in the mornings and so getting to my clinic for a nine o'clock start was becoming impossible. I started having lunch at my desk and then even while driving in my car to my next appointment, but there were still not enough hours in the day to fit in every patient.

One day in clinic I was shocked when a long-standing patient complained that I seemed distracted and wasn't giving my usual attention to her little boy's care. I had to admit this was quite true and so decided to acquiesce to the wishes of our nurses and receptionists and let them organise my schedules of clinic appointments, house calls, and visits to the wards. Nevertheless, I was seeing less and less of home and family. The girls had usually gone to bed when I got home and, at weekends, if we were to go out together for a walk or shopping, they often had to wait in a hot car outside the hospital as I rushed round my charges in the wards upstairs. I was becoming tired and dispirited. My family felt ignored and I was behaving in the same way as the over-worked businessmen that I counselled who were experiencing burn-out in their jobs, marital discord and undisciplined behaviour from their neglected children. A change was needed. I started to look round for another doctor to ease the work-load.

For any small organisation like ours with only four doctors it was quite a risk to take on a fifth person. Suddenly the outgoing expenses would be increased by 20% and to expect a similar increase in income as a result might be optimistic but the alternative was to start saying no to new patients.

David had been working at the Ruttonjee Sanatorium in Wanchai under Sister Gabriel. Sister Gabriel, an Irish nun, was a chest physician who came to Hong Kong at the height of the tuberculosis epidemic of the 1950s, teaming up with the famous Sister Aquinas for twenty years before her retirement in the mid-1970s. Sister Aquinas had pioneered the Asian studies of the wonder drug Streptomycin which had completely changed TB management and she was considered a world-authority because of this research; all done at the Ruttonjee. David, training as a physician, was very open to my suggestion that he join the practice for a year or two. However, I think my plan to work less did not really succeed as he immediately attracted his own legion of followers and was soon busy himself; but at least he was able to be on call on alternate nights and weekends taking out of hours calls.

With three girls needing to be with their dad at least occasionally, I was under family pressure to give up my 505 sailing. The class was dying anyway with ever fewer competitors in the races as the better sailors changed to the Flying Fifteen Class, a much more pedestrian boat that was to stage its World Championship in Hong Kong soon and naturally everyone wanted to join in on that. I found the FF rather dull and heavy after the 505 and anyway I had discovered something much more fun, windsurfing.

Our summer holiday that year was two weeks in Kauai sharing a house on a beach with some American friends and their children. Close to our rented house two young men with shaggy hair had some windsurfers to rent and were offering lessons. After a couple of sessions, I was hooked and as soon as we were back in Hong Kong I bought a second-hand board and was soon trying to sail it at Deep Water Bay. The family would all picnic on the beach while Dad repeatedly fell in the water, but gradually I found my balance and was soon sailing confidently around the bay. This was a much more family-friendly sport and the girls could splash in the sea or dig in the sand while Joan and I took it in turns to conquer the windsurfer. Windsurfing became almost an addiction and the board sat on the car roof most of the time. Any free moment I would launch from the nearest beach. I began to meet other enthusiasts and having successfully organised some competitions we decided to set up an Association. This would before long become WAHK [46], and forty years later, remains the body that runs windsurfing in Hong Kong.

I was well placed to publicise the sport as for the past two years I had been the yachting correspondent for the SCMP, the main English language newspaper, writing a column for the sports page every Tuesday under the alias of Seahawk. Seahawk reported on sailboat racing in the Harbour and offshore. Since I mentioned everybody in the articles it had a big readership among the yachties who, like anyone else, liked to see their names in print. I started to use it as propaganda for the sexy new sport of windsurfing. It worked, and soon we were organising ever larger regattas and even found some sponsorship; Marlboro cigarettes provided the money for a motor boat and its driver to organise the racing and provide safety cover. Within a year we held our first National Championship in Tai Tam Bay off Stanley Beach.

I admit that I was unhappy at accepting filthy lucre from the tobacco industry but theirs was the only finance being offered at the time. The Hong Kong government was encouraging but not yet materially supportive despite the fact that most competitors were local Chinese young people. I was elected president of the newly born WAHK and we set about raising more money and arranging competitions at Stanley and in Sai Kung in the eastern New Territories. We decided that properly trained instructors would be needed if the sport was to expand and so a young man called Steven came out from England for a few weeks to train our first cohort of local instructors. I decided to take some days off and do the course myself; but then disaster struck.

Bill, now in his early 60s, liked to play a not-too-gentle game of squash at the Jockey Club with his nephew Matthew once or twice a week. One afternoon Bill had a heart-attack and collapsed on court with chest pain.

---

[46] WAHK; Windsurfing Association of Hong Kong.

Matthew called an ambulance which rushed Bill to the Canossa. On arrival at the hospital Bill had a cardiac arrest. Thomas Lee, a paediatric friend of mine, resuscitated him and now Bill was in ICU. Sister Jo called me with the bad news and I cancelled everything and rushed to his bedside where treatment had been started by one of the cardiologists.

I spent the next 36 hours with him. He had an unstable heart rhythm that did not respond to the usual drugs being prescribed by the cardiologist and he kept arresting in ventricular fibrillation. Several times I had to give him a cardioversion shock and didn't dare to leave him for a minute. Eventually he stabilised and I could snatch some sleep and return to normal duties. Bill was to be off work for several weeks and, even when he returned, he lacked the powerful energy that had marked everything he had done before.

With the practice down a man there was no way that I was going windsurfing and Joan, initially reticent, took my place on the course and passed with flying colours. Little did we know then what a profound effect this would have on our lives for the next few years. Joan was enthused at the thought of using the new methods and planned to start teaching. She bought some boards and their equipment. From Australia she imported a dry-land simulator. Full of energy, she advertised in women's clubs' newsletters and the learners started coming. Her fees were not cheap but windsurfing was novel and trendy, while the learners were well-off people. Soon her books were full. Her working clothes were a swim-suit and a straw hat and her office was Stanley Beach.

She was very popular and only the least athletic failed to master the sport. For three years during the summer months when the sea was warm, she was on Stanley Beach several days a week. Her pupils recommended her to others. On Saturday lunchtimes her best friend Vanda would bring their family's junk with her two little girls and our three as well. They would anchor a few yards off the beach and Vanda would serve sandwiches and drinks to all comers including any of the pupils who were able to sail that far.

Joan Windsurfing

Vanda's husband, Stewart and I, if we had time from work, often swam out to join them. Joan's earnings mounted up and eventually became enough to pay the deposit on a London flat; so, we became property owners at last. After three summers teaching Joan decided to wind up her school in the autumn and sold the equipment. She had new ideas for a business.

With a friend she had conceived the idea for a service to assist the newcomers who were arriving in Hong Kong in ever increasing numbers. She christened the new enterprise 'Relocations Limited'. New arrivals came from many countries and, as many had hardly even left their home towns before, they were bemused by Hong Kong's ways, the language barrier and often just by home-sickness. Joan's plan was to take them in hand, find them a flat, a school for the children and, perhaps a maid. She arranged bank accounts, ID cards, driving licences and how to navigate the sometimes byzantine local bureaucracy. This saved time and money for the employers and human resources managers so Joan could charge good fees for her services. The social side was not forgotten and she organised get-togethers for wives and children until they had formed their own circles of friends. Starting out working from home they soon had to open an office and engage secretarial staff as human resource managers beat a path to their door.

Sometimes the bread-winner was not the husband and as his wife toiled in the bank or law-office the house-husband was also taken in hand by Joan's service. It became quite the thing for him to appear at the organised coffee-mornings and tennis matches.

CHAPTER 24. A GANGSTER, SOME WHITE-RUSSIANS AND AN OCEAN RACE.

With Bill unable to work, I was having to attend to a number of his regular patients as well. Among them I found some most interesting characters; perhaps not so much from a medical perspective but rather for the people they were and the lives that had brought them to Hong Kong.

Brett Downs was as thick-as-thieves with Bill and a fellow member of the Jockey club. Bill had famously won Brett's elderly gold Rolls Royce from him on a bet. With a shaved bullet head and a bull-neck above wide shoulders, Brett always reminded me of the New York gangsters in the Damon Runyon stories. Later I was startled to find out that that was exactly what he was. One evening Joan and I had been dining with George, a high-end antiques dealer, in his flat above his shop in Hollywood Road when George started talking about Brett, whom he also knew well. He produced a black paper-bound booklet, its cover emblazoned with the United States emblem of a heraldic eagle and the written legend 'Records of Congressional Hearings' and suggested I should find out about Brett's shady past.

It seemed that about ten or fifteen years earlier Brett had been the 'enforcer' for a group of 'businessmen' supplying gambling-machines and booze to American military servicemen's clubs throughout Asia, all the way from Vietnam and up to Japan. By guile, bribery and threats, their business had apparently exploited servicemen by selling inferior brands of beers and spirits to their clubs. They also rented out 'one-armed-bandits' and pinball-machines at inflated prices. After the fall of Saigon, the facts started coming out and some of the gang were sent to prison. Brett, who I was to find was named several times in the Congressional Report, had apparently escaped investigation by refusing to leave Hong Kong which had no extradition agreement at the time.

None of this, which happened years before I met Brett, changed my feelings about him; he was a thoroughly likeable man. He was very tough too and always tried to avoid giving me any trouble when he was unwell. Like all of his type, he liked to smoke, to drink and to gamble. His idea of a good weekend was to be out on his boat with a pack of cards, some good poker players and several bottles of bourbon.

Inevitably this lifestyle caught up with him and he had a huge heart attack one night. Typical for him, he suffered through the early hours before phoning me at first light. He told me in great detail how to find his apartment in the mid-levels, but as I drove up towards the address, there he was, standing on a street corner gesticulating. He was ashen, I dragged him into the car and rushed him to the Canossa where after a week or so he made a good recovery. The cardiologist decided that he needed coronary artery bypass surgery so I telephoned Michael de Bakey in Texas. Doctor de Bakey was a pioneering heart surgeon who had recently published papers on the excellent results he was achieving from surgery of the coronary arteries. He and Norman Shumway at Stanford were the first to perform this type of surgery that nowadays has become routine and can bring long survival for patients with heart disease. Although he recognised the severity of his illness, Brett absolutely refused to go to see either of these specialists in the US. It was only a few years later, after George's revelations, that I was to realise why, instead, he flew off to Manila where a local surgeon who had previously worked in de Bakey's unit, performed the by-pass. Brett's was the first such operation this doctor had ever done by himself. Brett lived for many years after, but this was the last time he ever strayed outside Hong Kong. He died, not from heart trouble but by electrocution while trying to fix an electric heater that was leaking water in his kitchen.

Another patient of Bill's was a White-Russian lady who had arrived in Hong Kong via Harbin and Shanghai, escaping first the Bolshevik revolution as a girl and again in 1949 from the Chinese Communists. During these adventures she had lost her husband, a Russian count, and when she came to Hong Kong she was working in a very smart shop in Central selling dresses from the great couture houses like Chanel and Balenciaga to the local tai-tais. She was a beautiful, charming and educated woman in her seventies. Her son had schizophrenia. He was a gentle soul and when on his medication, loved to talk about classical music when he visited me; but if he stopped the tablets, he quickly became violent and dangerous. I had stitched up his mother more than once after he had attacked her with a kitchen knife, his preferred weapon. Eventually a magistrate decided that enough was enough and had him incarcerated in Castle Peak Mental Hospital. His mother then became progressively more decrepit from worrying about him and had to give up her job. I visited her in her little flat in Pokfulam when she could no longer travel to the clinic, until a few months later all alone in the world, she died in her sleep.

There were a number of White-Russians living in Hong Kong. They had escaped by the skin of their teeth in 1919 across the wastes of Siberia to end up stuck in the northern Chinese city of Harbin, where as refugees they had survived as well as they could. Stories of young girls of noble birth working in brothels in the Treaty Ports are certainly true though I suspect any working-girl at the time would claim such distinction as an attraction to customers.

Elise and her husband, also White-Russians, had survived the Second World-War in a Japanese internment camp in Shanghai and subsequently stayed on there as race-horse trainers. They had to leave Shanghai in a hurry in 1949 for Hong Kong and set up again training the Jockey Club's horses in Happy Valley where they were moderately successful. Elise, now widowed and in her 80s, was living in an attractive old colonial style detached house near Stanley on a cliff overlooking Tai Tam Bay. She had a large garden and a shady patio open to refreshing sea breezes. Her daughter, Buzzy, was married to the headmaster of the Sea School nearby. One day Buzzy called me in after Elise had fallen out with her previous doctor of many years. From the start, we liked one another and, although old and immobile, she had a brisk mind and many tales to tell. Apart from listening to her stories, drinking tea made in a real samovar and writing her prescriptions, I had little to do, but I called in to see her monthly. Eventually her house was demolished to make way for the American Club which was to be built on its site. Elise moved into a beachside flat in Stanley and lived to a great old age. Buzzy and I remained friends, even after she herself retired to Scotland and lived there into her nineties.

Elise had defied her father in 1922 and sailed from China to San Francisco where she bought a car and drove it, single-handed, right across America to New York having many hair-raising adventures. From there she sailed back to Shanghai via the Panama Canal. She had been quite a girl. It would have been a daring adventure in the 1920s but perhaps not so great a challenge for a girl who had recently escaped from Russia at the height of a revolution.

As Bill recovered and returned to work, I started to take a more serious interest in offshore sailing and attended evening classes in navigation at the Polytechnic College in Kowloon. Hearing of my embryonic skills, a solicitor friend, Colin White, invited me to join his crew as navigator on *Wimera*, a locally built 35-foot sloop from the Cheoy Lee shipyard on Lantau Island. The plan was to compete in the China Sea Race to Manila at Easter. I spent hours on clear nights on our roof learning to use my new sextant to take star-sights. I pored over the charts and checked the radio channels on the short-wave set on the boat. We practised night sailing round Hong Kong and joined in the overnight races.

One dark night we were sailing in an area of rocky islets called, reassuringly, 'The Furies', scooting along with a good following wind. At the helm I was enjoying myself as *Wimera* sped across the sea when an unexpected wave jerked me off my feet. As I stumbled backwards, I managed to pull the steering wheel out of its mountings in the binnacle. I heard the chain that linked the cogs on the wheel spindle to the rudder fall with a rattle into the bowels of the boat. With no steering, *Wimera* promptly took command and veered off course. As the boat thrashed this way and that, Colin eventually managed to engage the emergency tiller and we were under control again, relieved that we had not struck any of the Furies. This was something we had to fix before we sailed for Manila, as we couldn't risk that happening again far out at sea.

We sailed back towards our moorings, tired by the night's activities. We had practised sail changes, we had retrieved a floating bucket several times in man-overboard drills, and we had set and doused the spinnaker. We were wet and cold so tea was brewed, but now our chef-to-be discovered he was sea-sick whenever he went down into the galley and so he had remained all night up on deck. I made bacon sandwiches for everyone except the cook who was too ill to even look at them. He obviously had to be replaced, and soon.

Wisps of mist were blowing off the land while the breeze faltered, and as we motored past the little light-tower by the whitewashed cliff-face of Tat Hong Point that marks the eastern entry to the Harbour, the fog closed in so we could hardly see the foredeck from the cockpit. We edged ourselves closer inshore and away from the shipping-channel where we could hear large ships sounding their horns. In shallow water we would be safe; so, when the echosounder gave us a 10-metre depth and we were nicely tucked into a bay we dropped anchor. We went below to sleep, leaving one of us to stay on watch in the cockpit as a precaution. We did not sleep long because the lookout heard a large boat moving nearby. As I emerged up the companionway, I too could hear an engine-room telegraph ringing and a large diesel engine changing revs as the bow-wave of some vessel approached. Out of the fog bank emerged a green painted tug-boat. Its bridge towered above us, radar aerial slowly rotating as it carefully manoeuvred alongside. The tug's skipper ran down a ladder and called out in Cantonese. It turned out he was lost in the fog and wanted to find the Harbour. He had located us on radar and wanted to know our position. I joined him and climbed up to the wheelhouse where he showed me his navigational chart, in fact it was a Tourist Association map showing the delights of the Colony, the very same one they handed out free to arriving visitors at Kai Tak airport. That was all he had for his job of towing enormous barges round local waters. I marked our position very roughly on his sketchy map with a Bic ballpoint and then off he went into the murk.

Yachtsmen and professional seamen do not often get on together very well. Hong Kong fishermen certainly seem quite aggressive towards pleasure boaters and will never give us an inch. In fact, this also seems true of fishermen all over the world. Perhaps somewhere in piscatorial lore there's a traditional verse that all young trawlermen must memorise; something like:

If a sailing yacht should cross your bow/ You might as well stop fishing now.

In the 1970s and 80s the local enforcement of safety rules at sea was very basic. This was especially true of showing lights at night. We yachties had to have all the right lights at the masthead and round the decks or face disqualification from our races. For local seamen, it seemed enough to hang a feeble hurricane lamp somewhere in the rigging. Red and green side lights were seldom seen. One night, we were quickly approaching a local junk chugging along in a haze of diesel exhaust displaying two red lights with a dim white one above them on his stern. That was not in our booklet of light patterns, but as we drew closer, I could see that it was actually two red candles on a little altar to Tin Hau in the stern window. Tugs with mile-long towlines often showed no lights at all.

Tin Hau is the goddess of the sea and allegedly worshipped by local seafarers. Temples dedicated to her are found in several fishing villages round the Colony with a large one on the beach near Junk Bay. Incense burns, mythical warriors glower, snakes and dragons entwine, gleaming brass buddhas with enormous bellies squat and grin. A man in a dirty T-shirt idles on a deck-chair beside the altar puffing a cigarette as the smoke from a dozen burning incense coils fogs the air. All very religious!

There is a tradition that a fishing boat, when leaving for deep waters, turns a few loops in the bay near the temple before heading out to sea. Since the Yacht Club's ocean-races starting-line was near Junk Bay, we yachties also followed the custom before setting off on the China Sea Race. An annual Junk Bay custom for Yacht Club members was a raft-up of yachts on the day of Tin Hau Festival but the event seemed more alcoholic than religious. An enormous pleasure junk that would have done old Kublai-Khan justice, is moored in the middle of a raft of yachts and smaller boats. Our commodore hosts the event and provides a cornucopia of drinks and 'small chow'. Afterwards, getting our boats back to the Yacht Club, several miles up the busy Harbour, was often a major challenge in our cheerful state.

Wimera did well in the race to Manila, winning her class and we, her crew, thoroughly enjoyed ourselves. There were several near disasters amongst the competing boats but nobody came to any harm in the end, although we were all amused by the increasingly desperate radio calls from a yacht called *Blue Dragon*. Her navigational plan seemed to be based on sailing southwards, then on reaching the latitude of Manila, turning left and sailing due east. A hundred miles or more after her eastward turn she still had failed to hit land. It turned out that she had somehow blundered too far eastwards and out through the Bashi Channel between Taiwan and Luzon, so had ended up sailing down the east coast of the Philippines. Her left turn eastwards took her further and further out into the empty Pacific Ocean. Some weeks later, I noticed *Blue Dragon* back on her mooring in Sai Kung so she had obviously managed to retrace her steps and was safely home.

China Sea Race on Wimera; with skipper Colin White.

We did experience one slightly scary incident on *Wimera*. On the final evening of the race, we picked up the regular flash of Corregidor lighthouse as we approached the coast. Relieved that my navigation had actually found Manila Bay, I went below to grab a quick sleep before the last leg of the course to the finishing line. The watch woke me in a panic an hour or so later. Corregidor light had vanished! One second the light was there but then it was not; the regular beat of the flashes ceased. No, they did not know what the compass-bearing was when it vanished. This was a problem as the light had a so-called 'occluded-sector' in which it could not be seen.

A boat in this sector would be in danger as it marked an area of shallows and reefs. Unfortunately, we just did not know whether or not we were in that sector. Although I thought it unlikely, for safety's sake there was no alternative but to turn 180 degrees and retrace our exact course until we knew we were out of the potential danger zone and could head for the finishing line a few miles further on into the Bay. These navigational problems are now things of the past as nowadays a glance at the GPS screen reveals your exact position, but in 1980 we had no such help. Once back on course we hurried towards the Manila Yacht Club finish line but had wasted an hour or so. It turned out that to save electricity, the Philippine authorities switched off their navigational lights at midnight. We had been quite safe all along. Yachties nowadays do not know they are born.

There is nothing you can do to help the dead. Olivia Manning.

In the early 1980s a new disease appeared that almost exclusively affected homosexual men, particularly the denizens of the famous men's bathhouses in San Francisco. It eventually became known as AIDS, but initially it was a frightening, little understood, incurable scourge. The diagnosis seemed to be a death sentence. Hong Kong was one of the centres of the fashion industry, which inevitably meant that in the very early days of the outbreak, we were amongst the first to see cases of infection, although we were unable then to identify the cause of their disease, which at the time was yet to be as much as reported in the medical journals.

A young American fashion designer was admitted to the Adventist Hospital with pneumonia. He was very ill and, not responding to treatment had to be ventilated in the intensive-care-unit. Pneumocystis carinii was diagnosed, an unusual yeast-like germ that affects patients with lowered immunity. He was cared for by several senior doctors, including some 'rock-stars' from the University. Even I was asked for my thoughts on the problem of his inability to fight this illness. One night, when things seemed at their worst for him, a vigil was held in the hospital chapel after which the tide turned and he gradually improved enough to be sent home, although still weak, thin and gaunt after his ordeal. Eventually he died back in New York. Many months later, a colleague who had also been involved in the case, showed me the latest copy of Time Magazine with the whole story of this new disease; so then of course we understood our young patient's diagnosis was due to a new virus that could suppress the immune system. AIDS may be still incurable but it can be managed and today those infected can lead long and useful lives.

One effect of the dread of AIDS was a minor sexual revolution. People were frightened of having sex even though condoms were thought to give complete protection. Certainly, the number of patients coming into the clinic with venereal diseases dropped dramatically.

Every autumn, the average age of the practice's patients increased quite a bit around the Lantern Festival, which by ancient tradition marked the end of the summer. As the weather cooled and the cicadas in the trees ceased their constant scream, there was a mass arrival of grandparents from all over the world to visit their expatriate families and to enjoy a holiday in the sun. These grey-hairs were to be seen everywhere: at the school gates, in restaurants and shops and in my clinic too. Forgotten prescription drugs, insect bites, coughs and colds, in-growing toenails and all the usual problems, but sometimes more serious conditions emerged. I can recall twice when old ladies did not wake up on their first morning in Hong Kong, presumably felled by pulmonary emboli after sitting immobile in aircraft seats for hours on the flight out. Especially sad was an old lady with a chest infection who had been admitted to hospital and was recovering well enough to expect to soon return to her family. One early morning, I went into her room and found her in bed, stone dead, but propped up on pillows with a hot breakfast steaming on a tray in front of her. She must have died an hour or more earlier but in the morning rush of starting the day, the ward staff had not noticed and her relatives were expected to arrive any minute! The tray was rushed away and, grabbing her ankles, I pulled her down the bed and drew the sheet over her face just before her daughter came into the room. That she had died was sufficiently distressing without her family realising that someone had served her breakfast without noticing she had passed on.

Deaths were quite uncommon among our patients because of the demographics of the population we cared for as among expatriates few old people chose to live in Hong Kong beyond retirement in those days.

Very occasionally a new-born baby died but the infant mortality rate in Hong Kong was, and still is, among the lowest of any nation, far below the UK or the United States. This is not entirely due to the excellence of the medical profession, though that is a factor. Chinese populations do in fact have a lower incidence of the congenital malformations which are a major cause of death in neonates. Nowadays, using high-definition ultrasound, abnormal foetuses can be detected early, allowing the pregnancy to be aborted. At the time I am writing about, occasionally babies did die soon after they were born with untreatable abnormalities of essential organs.

We also had to deal with Rhesus incompatibility, which caused deaths despite exchange blood-transfusions. After the introduction of Rhogam injections, already in use for several years at the time, rhesus-babies were no longer the fairly frequent problem they had been before the early 1970s.

Potter's Syndrome is a congenital condition that occurs when the kidneys fail to develop. In a healthy pregnancy, the baby's kidneys produce urine which goes to make up a large proportion of the amniotic fluid in which the baby floats. When the baby does not make urine there is less fluid and this is called Oligohydramnios. Before it is born, the foetus makes slight breathing movements, inhaling and exhaling the amniotic fluid but without enough fluid entering them the lungs don't develop properly. In addition, the pressure from the walls of the uterus, without the cushioning effect of fluid, impacts on the baby's face and limbs causing deformities. Before sensitive ultrasound machines were developed, the absence of the kidneys could not be reliably detected and Oligohydramnios was diagnosed when measurements of the growth in size of the uterus were unduly small. This was the situation when I was called to the labour ward to attend such a delivery. Apparently, nobody had warned the parents that anything might be amiss.

When the baby was delivered, he showed all the features of Potter's Syndrome. I knew that he could not survive without kidney- and probably lung- transplants, both extremely unlikely even in the best centres nowadays. The little boy did not start to breath on his own and I inserted an endotracheal tube but excessively high pressure had to be applied to achieve any ventilation. We continued trying to start his breathing while I talked to the parents, whom I had not met before. Together we decided that in the circumstances treatment was futile and that the baby should be allowed to die peacefully. After removing the breathing tube, I wrapped the baby in a shawl and gave him to his mother to cuddle as we watched his life fade away as he lay in her arms. We took a few Polaroid photos to help them remember him. There was no priest in the hospital and, although I'm a non-believer, I blessed and christened him with the sign of the cross on his forehead. I was quite surprised when I realised that I was actually shedding a few tears at the tragedy and the waste of a life

Occasionally an older baby succumbed to cot-death, but surprisingly only among the expatriates. These unexpected sudden deaths seemed not to happen in the Chinese population who liked to tightly swaddle their babies and sleep them lying supine whereas the western preference was to lay sleeping infants on their bellies. Sleeping posture was confirmed as a major cause of cot-death by a New Zealand study a few years later and nowadays babies everywhere sleep safely lying on their backs.

I occasionally had to go to the government mortuary which was situated near the Harbour in the western district of Kennedy Town. It was part of a complex that ironically included enormous waste disposal incinerators close by the mortuary entrance. Entering, it was like stepping back 50 years into a bygone age with marble dissection tables and open gutters on the floor. Huge modern fridges stood along the walls but the rest was Dickensian. I didn't dislike having to go there occasionally and I came to know the senior pathologist. Like many forensic pathologists, he had a gallows sense of humour and many salty tales about murders and other interesting ways of leaving this life; in fact he was a celebrated after-dinner speaker on his subject.

On one occasion, I had to visit the mortuary because of the sudden death of Andrew, a little boy with Down's Syndrome aged about 8 months. I was called to their flat by his mother whom I knew only slightly. Andrew was born in Scotland and had arrived in Hong Kong aged about 6 months but had not yet seen a local doctor. His mother was alone in their flat with Andrew whom she had found dead in his cot. She did not seem very upset by this and was in fact quite business-like. She hoped I had brought my Death Certificates with me ready to sign as she had already called the mortician. This briskness surprised me and rang alarm bells. Andrew's face was blue and there were tiny petechiae in the skin, a possible sign of suffocation. Sensibly, the law requires a doctor to have seen a patient during the last illness if he is to sign a death certificate. I could not do this for Andrew, so I had to call the coroner, or rather the police-officer who worked with him. The coroner's officer sent round his underlings and Andrew in due course arrived in the public mortuary where I had been summoned to observe the post-mortem.

The pathologist was able to show that Andrew had not been smothered and had died of natural causes probably related to a congenital heart abnormality. This made me feel a bit guilty about my suspicions towards his mother but on reflection I decided that being a bit suspicious would be expected of me since it is a physician's job to consider all possibilities.

Around this time, I was for the first time presented with a Living Will; it was brought to me by a one of Chris Pugh's maternity patients whose mother had arrived from Holland to live with her. The document, written in Dutch with an English translation, stated that Greta, who was 70 years old, had a form of Motor Neurone Disease which would result in her death within a few years. Since Greta feared the indignity of death from MND with its progressive paralysis, inability to swallow and loss of control, she did not want to be resuscitated or given any treatment such as intensive care or even antibiotics if they would lengthen her life. This was written in legalese and signed before a notary in Holland. After I had read it through, I agreed to meet Greta to discuss the matter. I kept a copy of 'the Will' which I showed to my Medical Defence Insurer who agreed on its legality and only suggested it might be signed again before a Hong Kong notary to be absolutely sure. There was no suggestion of euthanasia.

I met Greta at their flat overlooking Repulse Bay, its wide windows with a panoramic view of the beach and out towards the Chinese Lema Islands on the other side of the shipping channel. Greta was in a wheelchair and unable to speak. She was also unable to swallow, and drool ran down her chin to be constantly wiped away by her daughter. She had to be given her meals through a tube and had lost a lot of weight, in fact she looked like a little bird. A young woman in nursing uniform sat with us. Despite all this disability, Greta seemed cheerful and could indicate yes, no and maybe, by giving a little flap of her hand. I read through the document aloud with her and asked if each part was her wish. 'Yes,' she flapped and that was certainly good enough for me so I asked the nurse to write a few words on the back of my copy of the will as a witness to our conversation. I saw Greta from time to time until she eventually and inevitably developed a chest infection. We all discussed what to do. Greta refused antibiotics but did agree to a physiotherapist helping her to cough and breathe. Within two days, she lapsed into a coma and died the next day. It was what she had wished. She had died quietly and with grace at home among her family and had kept her dignity to the end.

I am often asked about euthanasia and I have to say that I do not approve of mercy-killing at all. I am convinced that doctors must never deliberately do any harm and no patient should ever have even a glimmer of suspicion that his physician might entertain lethal intentions towards him. There can be no grey area in this. On the other hand, 'to officiously strive to keep alive' is quite as bad as mercy-killing. When the time comes, a doctor must know when to stop and Living Wills lay down clearly the limits beyond which patients do not wish to go. Greta had drawn a line and we stuck by that.

In 1981, termination of pregnancy was legalised in Hong Kong, 14 years after this happened in Britain. This was welcomed by almost everyone although several of our private hospitals run by religious groups could not countenance abortions. The Matilda Hospital, having no religious connections became one of the few places a woman might seek an abortion. I worried about the operating theatre staff who had to witness these procedures repeatedly. To watch the bloody remains of what was almost certainly a healthy foetus gurgling through the transparent sucker tubing could be most upsetting. I am sure few, if any of the women and their doctors, took the matter lightly, but I was concerned for the staff, who had to stand by and later clean up after the procedures time and again. This was not what they had expected when they chose a career in healthcare.

Only gynaecologists were permitted to carry out abortions, but it was to ordinary doctors that women with unwanted pregnancies usually turned to initially and we had to help in the decision making, or counselling if you wish, and then refer them on to the specialist. For me, it was no different and I would never stand between a woman and her need for a termination. On the contrary, it was my policy to listen and let the woman tell me what she really desired. Some kept on with the pregnancy, but most did not. One of the first patients I referred for a termination was fifteen years old, a pretty red-haired Australian girl with adorable freckles. Andrea had been admitted to hospital at a weekend after some vomiting. Her parents, unwilling to disturb Bill or me, accepted the first doctor available. He considered either a kidney infection or a stomach ulcer as possible diagnoses. Over the next few days, Andrea was given several injections to relieve the vomiting as well as antibiotics and other medications. Worse, she had X-rays of her gall bladder, stomach and kidneys using contrast materials. Over thirty X-ray plates lay in her files.

The vomiting continued and her frustrated parents sacked the first doctor and asked me to see her. As I sat down with Andrea of course my first question was to discover the date of her last period. Predictably, it was a few weeks overdue. A sensible girl, she knew she was pregnant, though had not yet dared to tell her parents. She could not understand why the doctor had not considered pregnancy and she had therefore come to the conclusion that she must really have something seriously the matter with her. Of course, after all that medication and radiation, the pregnancy could not be safely allowed to continue and she chose to undergo a termination.

Even natural deaths bring extra problems when they occur abroad. Repatriating a dead body is a real rigmarole and very expensive. Travel insurance policies try to avoid paying for this service and so most people settle for cremation and then take the ashes back to their home countries. In Hong Kong, the funeral facilities are very modern with several large companies competing to provide the necessary services. Special among the Chinese community is that an auspicious date and time needs to be chosen for the various obsequies. The slots for cremations at the government crematoria may be booked several months in advance, long before most of the likely customers even feel ill. Like many other things in short supply in Hong Kong this, like taxi licences, has been turned into a racket which, as usual, is controlled by the Triads. 'Agencies' book all the auspicious time slots and so, if Granny is to be sent off on a lucky day then someone has to negotiate with the Triads and to secure a really lucky date, the family may end up paying out a lot of money. Expatriates on the whole are immune to such sophistry and are content to be cremated on any convenient day, but woe betide a local family that seemed to be saving money by choosing an unlucky time for their relative's funeral. They might never live it down.

This has been such a morbid chapter so far; I shall move on to something bustling with colour and life: how about a circus?

In the early 80s, the showman Jerry Cottle brought the big top to Hong Kong for a few months every year to escape the European winters. I suspect there was some form of tax break in this too. The circus was set up on a flat area at the Ocean Park in Shouson Hill. Tents went up, animals were kept in cages and the human performers lived in caravans. Posters were plastered all over town to announce the performances and everywhere children were beside themselves with delight and anticipation. Our three little girls were longing to see the acts and the animals. The smells and sounds of the circus pervaded Shouson Hill.

I was surprised to receive a phone call at the clinic from Mr. Cottle himself, asking whether I would mind driving round to Ocean Park to look at a baby chimpanzee with a cough? Suggesting he called a vet did me no good at all. His chimpanzees were attended by paediatricians as he claimed vets were fine for horses and lions but chimps needed proper care.

The tiny animal lay in his trainer's arms, big eyes alertly watching me as I approached. He had a metallic cough which sounded very similar to croup in a human baby and he also had the same 'catch' in his breathing. He had a slight fever too. Listening to his chest was complicated by scratchy noises from his body hair rubbing on the stethoscope diaphragm; but after a minute or so listening I could tune that out and hear his breath sounds properly. I ordered some simple treatments including steam inhalation but not antibiotics, as the infection was probably viral.

I called again the next day and everyone who could squeeze into the caravan had done so, including the lion-tamer and two clowns. All were delighted at the chimp's improvement and the little mite put his long arms round my neck and swung himself up onto my shoulders. I hoped this minor success with the chimp would not let me in for caring for a big cat or one of the elephants, but several performers of my own species did become patients at our clinic. After a few days, the chimp was well enough to come to the clinic where he delighted the children waiting there to see the doctor.

One of the clowns had a full-grown monkey which was part of his act, but this monkey disliked the third performer in their routine, a circus dwarf who had the classic features of achondroplasia. One day the monkey got loose and bit the dwarf rather badly on the arm. The victim and the clown came down to the clinic in Repulse Bay, disappointingly wearing normal clothes. I cleaned and dressed the wound and administered an injection. However, the clown was not finished and back in the waiting room and with an audience of patients, he put on a worried look, asking whether we should worry that the monkey might catch some disease from biting a dwarf. The waiting children all thought this very funny.

It was not the only monkey bite I had to deal with. The second time it was much more serious and it happened at a pet-shop near the clinic. Next to the sandy Repulse Bay beach, under some trees, were several ramshackle retail stalls including this pet-shop. The lady owner, Mrs Kai, kept a large monkey chained in a tree at one end of the shop near the till. The rest of the area was crowded with aquaria, cages of gerbils and hamsters and buckets of terrapins. The place smelt just as you would expect and children loved it, mine included. Mrs Kai had a boy-friend, a taxi driver of whom the monkey was very jealous. One afternoon the monkey managed waylay the taxi driver and bite his hand. It was a very deep and vicious cut which severed several tendons and required a long stay at Queen Mary Hospital. I never heard what happened to the boy-friend but the monkey was still around several years later.

CHAPTER 26. SASSOON ROAD.

We had now lived in Hong Kong for nearly a decade, it was 1983, I was turning 40, and was worried that my medical knowledge was getting out of date. My medical registration in the UK also required updating after some new legislation. After some discussion with the Royal College in London, I realised that I needed to go on a refresher course in order to remain on the register as a GP. A Hampshire general practice agreed to be my host for the necessary three months. I rented a family house nearby, and bought an old banger of a Saab to get around in. Joan and I, plus the three girls, spent an idyllic summer in rural England. The two older girls enrolled in the local village school and I worked as a GP in the nearby town under the eye of a GP trainer. I was designated 'trainee' which suitably put me in my place. I went to seminars, attended ward rounds at the local general hospital and also spent a week in an inner-city practice in faraway Liverpool where we all could stay with Joan's family. I learnt a lot. Bubbling with new ideas and feeling totally refreshed, we flew back home in September. Hong Kong had indeed become our home.

Joan was keen to get back to her nascent business now called 'Relocations'. She was taking on staff and writing up a handbook to be presented to her clients on arrival. Discussions on copyrighting this dossier were under way with her solicitor, Robin Bridge. Robin was an old China Hand and was among those severely injured in the Kotewall Road disaster many years earlier when he was a junior solicitor.

The practice seemed not to have missed my presence for three months so I rushed round reminding other doctors of my existence. It was a bad time for Hong Kong as political discussions in London and Beijing had opened up the problems of the 1997 handover to China. Previously this was the 'elephant in the room' that we all chose to ignore. But ignore it we no longer could, as it was barely 14 years away, and had already prompted an exodus of many good people, including some doctors whom I knew well and had relied on. They were naturally worried about the future for their families and anxious to see them established in new homes away from the Chinese Government they had known too well before they escaped to Hong Kong. Vancouver, Sydney and London were the beneficiaries from this drain of brains and money from the Colony. One of my paediatric friends became an 'Astronaut', the term used for the breadwinner who remains in Hong Kong to work and has to keep flying in and out to be with his family in their new home abroad.

The general nervousness spilled over into the financial world and there was a run on the Hong Kong dollar. The government managed to defeat the raiders by raising interest rates massively and then they pegged the local currency to the US dollar. This has kept confidence in the local currency ever since. The local stock exchange index, the Hang Seng Index, which had been around 1400, plunged to half its value and did not recover to earlier values for another three years

Meanwhile the British Prime Minister, Margaret Thatcher, tried to tell the Chinese Government exactly what they should do about Hong Kong and received a rude riposte. 1997 was not negotiable. Of course, everything in life is negotiable, and the next years were spent in often acrimonious talks on the matter. Meanwhile we mere mortals got on with our lives as best we could, enjoying the Hong Kong lifestyle, working hard and playing hard.

Our lease at Ann Gardens was soon ending and our landlord, obviously ignorant of the sorry state of the housing market right then, was insisting on a large increase in rent and so we had started looking for another flat to move into.

For the past few years Hong Kong had been agog at the antics of George Tan and his company which rejoiced in the dreadful name 'Carrian'. There were rumours about Carrian's vast financial backing though nobody seemed to know where it came from. The source of the money might have been Indonesian billionaires or perhaps Imelda Marcos. A shady banker from Malaysia named Osman, seemed to have arranged a vast loan to Carrian without any collateral. Tan bought Gammon House, where we had our office, for HK$ 1 billion and resold it within a year reaping an apparent profit of 60%. At the same time, he was investing heavily in real estate all over the place, plus buying a bus company and a pest control operation. Carrian could be seen everywhere, Tan was on every front page. All the banks wanted to lend to him and the usual precautions seemed to have been dispensed with.

One of my patients, who worked for a very conservative London based bank which had not yet lent to Carrian, was being courted by Tan. At a cocktail party my patient, let's call him Tom, admired Tan's fancy wristwatch. (This is a convenient conversational gambit when trying to cross the cultural and language divide in Asian business). Tom was alarmed next morning when George Tan's messenger arrived at his office with a gift of two similar watches, one a lady-size for his wife. Tom asked head office and was told it was acceptable to keep the gifts. Next, Tom received an invitation to a family dinner at Tan's huge and garish home on the Peak. After the meal, Tom was saying goodbye to Tan and was about to call a taxi when George offered him a lift in the Jaguar parked in the driveway. 'No, you drive,' said George who didn't have a licence himself, giving Tom the keys. 'Anyway, the car's yours' added Tan. 'Enjoy it'. However, a car was a step too far and Tom was ordered to sever all connections with Carrian.

Even I, as the president of the windsurfers, was given money by the great George Tan. At a ceremony in his penthouse office, he presented me with a cheque to sponsor a regatta in Stanley and to buy equipment to get young people started in the sport. The cheque did not bounce and the money helped advance our sport.

Then of course, all the wheels came off. Property prices plunged and questions needed answering as Carrian's share price slumped. The body of a Malaysian financial investigator was found on a remote hillside and a prominent Hong Kong law-firm partner drowned in his own swimming-pool, weighted down with a manhole-cover tied round his neck. Tan and Osman vanished. The money had evaporated and creditors went unpaid. Half built apartments stood empty all over the Territory, and many a small construction company went bust. The ICAC sprang into action and several minor players pleaded guilty and were jailed. Bankers were disgraced and sacked. Tan and Osman were eventually jailed despite the prolonged efforts of their defence lawyers.

For the Howards, the crashing house prices were a godsend: we could suddenly afford to climb onto the housing ladder. One of the big banks was ruefully holding a portfolio of Carrian's failed projects having believed they were trustworthy debtors. They were anxious to sell and we were able to buy a townhouse in Sassoon Road from them. We loved the house as soon as we saw it. It was perfect with sea-views all round, enough bedrooms for everyone and a small garden and terrace. It was only on our second walk round the house that we realised it had no kitchen. There was a big void under the house where it was supposed to be, but no kitchen. Nevertheless, we got a mortgage, employed a builder and before long moved in. We lived there for 14 years until the girls had all departed for university. After several happy years in Shouson Hill, we were able to wave goodbye to the dreadful polisher family.

Sassoon Road, A leafy cul-de-sac.

While it may have seemed lunacy to buy property in the negative economic and political climate that we were experiencing in 1983, for the Howards it made total sense. Prices were low and unlikely to fall further and also for the last nine years we had been spending a large part of my income paying out rent to landlords. This was an opportunity to build up some capital in a hopefully appreciating asset and, more importantly, knowing that we were to be in the Colony for the long term, we could reduce our exposure to the inevitable future inflation in rental prices. That is not to say that my heart was not in my mouth when I signed the purchase agreement and mortgage documents, knowing that interest rates had just hit 17%. In retrospect it was the best decision Joan and I ever made, apart from when we decided to marry all those years before in Liverpool.

The move seemed to please everyone especially Sascha's cat, Middy. She had never liked to venture outside the Ann Gardens flat, only occasionally looking out of the front door, taking a sniff, then rushing back inside to safety. Arriving in Sassoon Road, she went straight out to the garden where she met another cat and disappeared for several days; eventually bringing her new friend back with her, though we chased him away. For Spitz though, it was not such a good move. In Shouson Hill he had been cock-of-the-walk and most new litters of puppies in the area had the Spitz look about them. At Sassoon Road, the local dogs were bigger and rougher, and poor Spitz kept returning from his amorous adventures bleeding and with torn ears. He would lie low, forlorn in his basket for a few days then venture out to return badly mauled again. The only answer was a visit to the vet, and Spitz, now well into his second decade, unknowingly gave up his love-life permanently.

Although the children loved the new house, we were no longer within walking distance of the Country Club as we had been in Shouson Hill. On summer days, after school, all three girls had liked to head for the club's pool where the two younger ones were champion swimmers in their age groups. Diane, who had become a sophisticated teenager, was far too cool to compete seriously. After all, she was a top-class tennis player in her age-group. The logistics of driving the children to and from the club with two working parents, were becoming difficult and so the problem was solved by asking Prestige Pools to dig us a swimming pool in the back-garden area.

It was an instant success and the house was always full of youngsters from school and nearby flats. Joan and I got to know many of these kids and were able to take part in our children's social lives.

The house was situated near the end of Sassoon Road which was effectively a quiet cul-de-sac where children could safely ride bikes and skate-boards, or just hang out together. Spitz liked the freedom to wander there as well. He liked to meet the girls off the school bus which stopped at the end of the road. Often the neighbours' children saved food from their lunch boxes for him, but he would have met them all anyway, being such a sociable canine.

We had several interesting neighbours in our compound. The couple at House-number-one had a mynah bird whose cage hung in a tree by their front gate. This bird drove his owners crazy by constantly imitating the ring of their phone. It had Spitz fooled too after it learned my whistle for calling him back home.

There were some feral cats living in the undergrowth nearby which looked underfed and ill, and were generally quite vicious if cornered. I found one lying apparently injured in our car-port. I assumed Spitz must have mauled it as it was very sick and seemed to be suffering from some internal injury. I put it in a cat-box and drove down to the RSPCA where they undertook to humanely 'put it down'. As I arrived home from this mission, the children crowded round me, clearly upset. The cat in question was very old and actually belonged to a neighbour. It was being treated by the vet for a quite serious illness, so Spitz was innocent and had not touched it at all. I rang the RSPCA but they had already sent the cat to his maker, so I had to explain my actions to his loving owner. While very unhappy about it, she accepted my explanation and I think we remained on good terms.

The two younger girls took up riding. The Jockey Club ran a stable of ponies and horses near Pokfulam Reservoir where Sascha and Julia would spend much of their free time. After some lessons, they were soon show-jumping and competing in dressage. In dressage, the rider has to be immaculately turned out herself, as well as the pony. Julia, then a tomboy who normally only wore shorts and old tee-shirts, turned out looking like Princess Anne, even wearing a hairnet and giving her boots a polish.

Fathers' day at the Jockey Club, in a minute
I shall have fallen off.

The ponies, with names like Browniecake, Tiffany and Tonto, were wilful creatures and turned disobedience into an art-form, but the girls learned how to keep them in line with a few good kicks and a flick from a riding whip. At Chinese New Year, the mafoos had their annual leave and the children were put in charge of mucking out and feeding the horses for a few days. The ponies were never so immaculately groomed as when this army of little girls took over with curry combs, Stockholm tar for the hooves and saddle soap for the tack. Manes and tails were braided and plaited, and the mucking out was done lovingly. I lived in fear of the father-and-daughter events when I was expected to ride round the ring on some enormous beast, inevitably falling off even before attempting a jump.

Living further from Stanley Beach, we were windsurfing less. The girls liked to stay at home and use their own pool and I was becoming more interested in sailing larger boats. One of Joan's windsurfing pupils, Sidney, and her husband Lynn, had become great friends. He was an interesting man, extremely tall; he had been a fighter-pilot in the US Air-Force before a career change into banking. He planned to race his boat *Innisfree*, another Cheoy Lee 35 like *Wimera,* in the China Sea Race and he asked me to join the crew. I was to be navigator again. We had a good crew but the light winds didn't suit our rather heavy boat and we were well back in the points.

China Sea Race, taking a sun-sight.

It is about 650 miles to Manila from Hong Kong. The first third of the route across the continental shelf is usually windy with choppy seas, then the winds drop and are fickle for a while until you reach the Philippine coast where the daily sea-breeze powers the boats south to Manila Bay. That year was no exception and we had a wet, cold ride well reefed down for 2 days. Waking on the 3rd morning, I came on deck to glorious sunshine. The flat sea was a brilliant blue but we were almost stationary. We hoisted the spinnaker and spent the hours catching zephyrs here and there. The four-hour watches on and off were exhausting in the sultry heat and down below without air movement, it was too hot to sleep. After dark a breeze sprang up, and with the spinnaker pulling hard, we tore across the completely smooth sea surface, our wake fizzing like phosphorescent champagne under the hull. Towing a fishing line, we caught a nice yellow-fin tuna which I, being the ship's surgeon, cut into strips and so we had fresh sushi.

After our arrival I enjoyed a lazy day at Manila Yacht Club, dozing and drinking ice-cold rum and calamansi juice, the local specialty, before taking the earliest available flight back home. Lynne made a longer cruise exploring nearby harbours and islands before sailing back to Hong Kong, but I had to get back to my patients and my family.

Shortly after my return from the Philippines, we were invited by an American friend to celebrate the visit of his brother Paul Theroux the author, for an evening and dinner out on a junk. Paul had spent several months in China, exploring the vast railway network all over the country in the last days of steam locomotion there. He had fallen quite ill and had lost a lot of weight but was recovering and planned to go back north to finish his book which we all discussed long into the night as we sat at the table after the dishes had been cleared. The conversation later moved onto the rapid changes happening in the PRC so soon after Mao's death. A year or so later 'Riding the Iron Rooster' was published and we rushed to buy it. It remains a firm favourite read for us both.

More work was coming to the practice and I decided to give up my teaching rounds at Queen Mary Hospital. I would continue with some contact with medical students as the Chinese University in Shatin had asked me to be a tutor in General Practice. Now, only one student at a time came to sit with me during my clinics, where I let them examine patients and take histories. It seemed that most of the patients enjoyed their presence, though a few made sour comments. As I've already said, teaching improves the teacher, especially when a student's questions provoke him to think closely about what we are doing for a patient. Students seemed surprised that we were able to do minor-surgery in the clinic. I liked to remove skin cysts, diathermise warts and keratoses and perform the Zadek operation on in-growing toenails. It saved the patient a lot of time (and money too) if we did these in the office rather than taking them up to the hospital. It was good for the nursing staff to keep their hand in by assisting and managing the instruments, sterilising them in the autoclave and so on. Some patients also found it hard to accept that a general doctor should perform minor procedures like cutting and stitching and Americans particularly thought this odd. A young lad had cut his hand between his thumb and forefinger, a superficial wound about 5 cm long, a clean cut from a kitchen knife. We were at the hospital out-patients' room and I had prepared to stitch up the cut under local anaesthetic, as it only needed 3 or 4 sutures. We were ready to start, when his father arrived and insisted that a specialist surgeon do the task. Not wanting to be awkward, I readily agreed and found a willing plastic surgeon. The boy had been eating his afternoon tea when the accident happened and so the surgeon insisted on 4 hours delay before a general anaesthetic could be given so the boy was admitted to a private room to wait. I heard later that the surgery was performed in the main operating theatre late that night, and the patient had to stay in bed for forty-eight hours afterwards, with his hand immobile and elevated above the bed. The final bill for the surgeon, anaesthetist, stay in a room for 2 days and so on had amounted to HK$35,000. If I had been left to treat him, he would have been home that afternoon with a bill for about $2,000.

In the autumn, Joan and I decided to take a short break and visit India. We wanted to see the Himalayas and I wanted to ride on the 'toy train' up to Darjeeling that had caught my imagination as a schoolboy reading *The Children's Wonder-Book of Engineering*. Our travel agent secured our reservations, a sleeper train ticket from Calcutta to New Jalpaiguri and the tickets for the Toy-Train that zig-zags up the mountains among high tea plantations to Darjeeling itself. We were to stay at the old colonial-style Hotel Windermere, 'with a coal-burning fireplace in every room'.

The *Lonely Planet Guide* told us that the old and atmospheric Grand Eastern Hotel in Calcutta was the best place to stay on our first night in India. We were shocked to find the place completely run-down with peeling paint in bare rooms with ancient beds. Joan nearly had a fit when we were taken to our room and she insisted that we be given a better one. This caused consternation as it turned out that we already had the best room in the place. We survived the night and made for the railway station to catch the sleeper up-country. Of course, we had to bribe an attendant to get the compartment and beds we had already paid for and during the argument, we encountered Tony Ohja, a Bengali businessman with a shining bald head who was accompanying his young son on his way to a boarding school in Darjeeling. After an evening drinking our bottle of dry sherry and his bottle of Chivas Regal as the train rattled northwards, we had become great friends. But then he brought bad news: there was a general strike at New Jalpaiguri, a 'bund' as he called it. The strikers had vandalised the Toy-Train track and it had not been running for over a year. They were now laying siege to the New Jalpaiguri railway station. Tony assured us it would all work out as he knew the police-chief there.

In the morning we stepped out of the train and were immediately told by the staff to get back in again and return immediately to Calcutta. It was too risky to stay because of a full-blown riot in the station yard. Tony rushed around to find his policeman friend and in no time we were in the back of a police Land-Rover speeding through the rioters. They banged on the sides of the car, rocking it on its springs. Joan and I wondered where we were heading for now. Tony of course had yet another friend, a tea-garden manager with whom he intended we should stay until things quietened down a bit. So we four, plus a young mother with her young child also headed for boarding-school, stayed for two nights at the Hansquar Tea Garden. It was a heavenly place with wide views of the high mountains and a clear blue sky. We walked among the tea bushes and watched the women picking tea which was then dried and processed. Arun Banerji, the manager, and his wife were perfect hosts as we shared their wonderful food. Eventually they found us a taxi that we hoped would take us all by the back roads up to Darjeeling.

After an adventurous drive at the end of which the rioters tried to push our taxi over a cliff, we arrived at a road block. The strikers manning it took a dislike to our rather aggressive and rude taxi driver, as had I. However, they allowed us passengers to proceed to Darjeeling by bus. So we clambered aboard the rickety conveyance standing nearby and waiting to leave. It was already overflowing with passengers and luggage; a monkey on a leash sat on its roof. Some young men kindly climbed on the roof with the monkey to give us their seats. An old man asleep across the back bench was awoken, and that made another seat. I sat on someone's suitcase and off we went up the steep incline of the Cart Road. It was already dusk when we arrived in Darjeeling's main square and Joan and I were about to set off in search of the Windermere Hotel when Tony recognised some men dressed in dark suits walking nearby. 'These are doctors, you have to meet them'. Tony really did know everyone.

The doctors said they could not stop to chat as they were on their way to express their condolences to a colleague whose mother had just died. Then, hearing I was a medic too, they urged us to accompany them. 'You should come with us and meet him'. Despite our protests that this was surely inappropriate, we were swept along with them to a large wooden house in a dusty compound. A robed Buddhist monk was at the front door and showed us in, giving us cups of tea and little cakes. We sat down and listened as the bereaved doctor gave us a blow-by-blow account of his mother's last illness at Beth Israel Hospital in New York and her death from pancreatic cancer. We were shown the X-rays and the scans, the blood tests, the operation notes; it was a masterful presentation that would have been worthy of publication in the New England Journal of Medicine. There was clearly something special about this Doctor Tampa Thondup whom I was meeting for the first, but not the last, time.

He welcomed us to Darjeeling, and when I mentioned my name, he said that he had already recognised me. Then I recalled that a year or two earlier at Kai Tak airport, I had sat next to a man carrying an ECG machine with a green case among his hand baggage; both of us were waiting for delayed flights. Hearing I was a doctor, he had asked my opinion about this machine which was intended for his brother's practice in India. Tampa pointed to a shelf on which the distinctive green ECG was standing.

Tampa asked where we had booked to stay that night and then insisted that we should be his guests. He would hear nothing against it; he would show us the sights and look after us. And that's what we did. I have never spent a colder night in my life. Joan and I wore all our clothes and clung together, shivering in one of the single beds; we even put the carpet from the floor on top of us as an extra cover. I thought wistfully of the Windermere's 'fire-places in every room'. Tampa had us out of bed before first light and into his Range-Rover to drive and see the dawn rise over

Kanchenjunga. It was well worth it and as the morning mist cleared away, we looked far up into the heights as the sun's first rays picked out the world's highest summits with Everest in the distance, higher than all the rest.

On one wall of Tampa's living room was a large photo of the Potala Palace in Lhasa. As we ate a delicious and spicy breakfast, he told us that this it had been the view from the garden of his childhood home that he had left in 1959 when the Chinese Army had occupied Tibet. His uncle was none other than the Dalai Lama himself and the whole family had fled across the mountains pursued by Chinese soldiers to become refugees in India. He had eventually trained as a doctor in Ireland and in the States and now he was working as a GP and paediatrician among the Tibetan community around Darjeeling. It turned out that it was his brother whom I had met in Hong Kong. He owned both a printing company and a noodle factory nearby. Sadly, we had to get back to Calcutta to catch our flight home and so we said goodbye to Tampa, promising to see him again, perhaps in Hong Kong.

With Tampa in Hong Kong.

Tampa did eventually come to Hong Kong where he worked at Queen Mary Hospital for a year, then I was able to sponsor him for a work-permit and he joined our practice. His calm and kind demeanour, arising no doubt from his Buddhist faith, made him immediately popular with everyone. He

216

married his girl-friend from Darjeeling and stayed with the practice for several happy years before leaving to set up on his own. Being able to alternate out of hours duties with Tampa made my life much easier; a good thing since I had become the practice's managing partner as Bill began to wind down and start thinking of retirement.

CHAPTER 27. XINJIANG.

The practice had in twelve years seriously grown from the two-man outfit of 1974 into an organisation with six doctors, a laboratory and three clinics. We employed a financial manager as we now had to negotiate with insurance companies and human-resources managers on a daily basis. I had to make time to keep an eye on all this and needed a secretary/typist to help me. This was made easier by the arrival of Tampa as I could now delegate some of my clinical work to him. Our first-floor clinic, up those dreadful stairs in Repulse Bay, had become too small to cope with the numbers we were seeing. It really required completely remodelling but Linda, our feisty hairdresser neighbour, had heard that there was space for rent in the nearby shopping centre above the Dairy Farm supermarket on Beach Road. So, after all those years, we moved to a brand-new purpose designed clinic with lots more room. An Australian GP, Anne Spooner, joined the practice to work full time in the new premises where I worked part-time and the other specialists had sessions once or twice a week. Linda soon moved too and set up her rather scruffy but lively salon across the corridor from us. Anne, who came from Brisbane, was immediately popular for her knowledge and her assiduity. Nothing was too much trouble and of course she had an immediate rapport with fellow Australians.

With Anne Spooner.

With more space than we needed in the new clinic, we started to sublet one of the consulting rooms to other professionals. The first to take this up was an optometrist who examined the vision of all our new-borns and also the 5-year-olds before they started primary school. Soon after, Helen, an American psychologist specialising in marital and sexual problems, rented space from us for a few hours a week. These two ladies were a great success and their expertise brought many new faces into the practice. One of Helen's patients, a French woman, called me one evening in great distress. With her children, she had just returned from a couple of months leave in her home country and her husband had failed to meet her at Kai Tak Airport as she had expected. Worse, when she arrived at their apartment in Repulse Bay, her key had not fitted the front door and she found someone else was living there. Her husband had completely disappeared and to add to her problems had blocked her credit cards so she was unable to stay at a hotel. A neighbour had taken in all the family so their accommodation was temporarily sorted out. As her English was not very good, I undertook to ring their employer's office and found that the husband had left the company and nobody knew where he was. The French consular service had to provide financial help to repatriate them to Europe. Clearly Helen's marital advice hadn't worked in this case.

Despite the pressure of my work, when the Friends of the Art Museum proposed a two-week trip to the far west of China, Joan and I were keen to seize this opportunity as it was an area that was normally almost impossible to visit because of security concerns. Not only was China's nuclear programme centred there but the local population were restive and resistant to Beijing's modernisation plans for the region. The Uyghurs are a separate ethnic group, descended from the peoples of Central Asia, they speak a Turkic language and are followers of Mohammed. Tourism was discouraged and information about the area was hard to find. This was where we planned to visit.

Accompanied by Chinese archaeologists and scholars, the Friends had special access to the back rooms in museums and to recently opened archaeology digs. Joan and I had already been on brief visits to Xian to see the terra-cotta warriors and the other great tomb sites that were being excavated. On one visit with the Friends we had been allowed into a recently discovered site, nowadays called the 'the dolls-house tomb', where we were welcomed with green tea and small cakes by the archaeologist in charge and then taken underground to see ranks of 2-foot-high figures emerging from the earth, where they had been buried for many hundreds of years protecting the Emperor Jingdi in the afterlife. Only a couple of pits had been opened so far and workers were clearing away debris so rows and rows of these little men's heads peeped above the ground. They estimated, from their preliminary surveys, that there were perhaps ten thousand of these little soldiers in outlying pits. Originally, they had been clothed and had articulated upper-limbs made of wood. Time had stripped them naked and rotted away their arms. In contrast to their much larger and very soldierly neighbours in the better-known pits of the Emperor Qin's tomb, they evinced a peaceful charm, they were indeed like dolls.

Now in 1988, with the Friends group, we intended to penetrate much further west, starting in Gansu province and then going on to Xinjiang. Far out in the desert among vast sand dunes, we were allowed into the Mogao caves in Dunhuang which are nowadays closed to all but the most serious scholars. They contain great libraries of books, scrolls and papers dating back many hundreds of years and preserved by the extreme aridity of the desert. In the early twentieth-century, they were only guarded by an ignorant and probably venal monk, who stood aside while explorers like Aurel Stein plundered them, carting their booty back to the great museums in Europe. The caves are still stacked with documents but they deteriorated quicky from contact with the tourist masses and so have been sealed up again.

Nearby we climbed up the vast sand dunes, arriving hot and exhausted at the top where one of those tiny ancient ladies you find in all the most unlikely places in Asia, sat under a sunshade beside a huge esky filled with ice and cold drinks plying her trade and charging extortionate prices. From our perch high above the desert, we looked down onto the perfect crescent shaped lake glinting in the vast arid wilderness.

Xinjiang, with Joan visiting a ruined city near Kashgar.

Driving on to Heavenly Lake in the Tianshan mountains, our next destination, we had to wait as Kazak nomads on horseback, wearing traditional dress, drove their herds of sheep and horses down the only road to their winter pastures. Great camels, laden with the traditional canvas yurts, plodded along behind the flocks, dogs barking round their heels, amid grubby children getting into all sorts of mischief.

In the Kazak city of Turfan, built 150 metres below sea level, we found a green oasis where vineyards and farms are sustained by an ancient network of underground irrigation tunnels which carry snowmelt from nearby hills under miles of desert to the town. These tunnels, the life blood of Turfan, need constant attention to remove the accumulations of debris that constantly threaten to block them. As we crossed the empty desert in our minibus, every few hundred metres we could see the tops of the access shafts surrounded by low walls. Beside them often stood a wooden derrick with a lone donkey hauling up loaded baskets of stones and gravel that his master was clearing from the tunnel beneath. We climbed down into one tunnel beside its rushing water and enjoyed the icy cold air ten metres down below the burning desert.

Next stop was Kashgar, which needed a special permit for a visit. Our group was rushed past the control point but then there was a delay and we milled round in the arrivals area of the airfield. A policeman, looking very cross, started to yell at us, 'Who is Thomas?' We quaked in our shoes feeling like naughty schoolchildren being interrogated by the headmaster. Finally, Thomas stepped forward with his companion. They were BBC reporters apparently sneaking into the closed-area by pretending to be tourists. They were rather roughly manhandled into a police van and not knowing what was happening to them, we were all herded into our bus worrying about their fate in the hands of the ferocious policemen.

We were accommodated in a scruffy hotel near the old British Embassy once the home of the explorer and diplomat Francis Younghusband and now just a rundown neglected building, a relic of the Great Game when the Russians and the Brits spied on one another and vied for influence on the roof of the world. The larger Russian embassy was at the time being converted into a smart new hotel: 'smart' by Xinjiang standards that is.

The specialty of our hotel chef was the local fat-tail lamb. In fact, the menu only offered lamb and melons. The nearby city of Hami, an oasis famous for melon growing, was nearby and it was melon season. So we had melon to start, followed by a whole lamb with little lightbulbs glowing in its eye-sockets. It had been slow cooked and it melted in the mouth. It was absolutely delicious. And then as we wiped mutton grease off our fingers and faces dessert arrived – melon of course. We were offered Hami beer with the meal, and discovered too late that, like everything else from that town, it tasted of melons. As we finished our meal, who should wander by our table but Thomas and his BBC colleague. They sat down at a table nearby accompanied by two uniformed policemen all enjoying a few beers. I went over to check that they were all right and they introduced us to the two Chinese officers. They had been released after an hour or so, having been admonished for arriving with the wrong papers. After that the policemen had taken them on a quick guided tour round Kashgar as they would have to catch the morning plane out. Now they and their captors were treating themselves to a slap-up dinner. The BBC, their employers, made their 'arrest' the lead item on the World Service news that night but they seemed to me to be having a fine time of their detention.

Food was often something of a problem when travelling in remote places. The areas we often visited were very poor and only beginning to recover from the economic stringencies of the Mao years, so meals could be very basic. We certainly never saw any of the fancy dishes served in Hong Kong's Chinese eateries. There was always plenty of rice but sometimes it was grey and gritty, vegetables were often stringy and overcooked, but this was the price of visiting this fascinating place at a fascinating time. Nevertheless, we were often hungry. At one hotel where we had been given a breakfast of nothing but bread and rice, we observed that a French group was eating hard-boiled-eggs. No, our party was not to be given any of these the waiter said, so we raided the kitchen in force and were quickly chased out by an angry chef, but not before we had stolen a bucket of nice hot boiled-eggs. A few years later, Joan visited the grasslands of Mongolia where the main staple seemed to be grass soup. Occasionally a lamb bone, stripped of all flesh, floated in the green liquid but with only chopsticks, nobody knew how to eat it.

The great attraction at Kashgar was the weekly horse and camel market. On Saturday morning the roads approaching the market became jammed with the little donkey carts the Uyghurs used for transport. Everyone crying 'push, push' in vain attempts to clear their road ahead. We elbowed through the throng, everyone was excited and happy on their way to the event of the week. Youths and young men dashed their ponies across the maidan showing off to the brightly dressed young women who mocked them loudly for their childish bravado. The smell of the animals, the human crowds and the smoke from the kebab stalls in the dry still air still lingers in my memory of this amazing place which was almost cut off from the world and could be visited only by a determined few.

Inside the town, the markets sold clothing, fabrics, jade jewellery, and electronics. Stalls displayed elaborately carved and fretworked wooden furniture. Everything was gaudy and brightly coloured in contrast to the unrelieved greyness of the desert. Many of the stallholders were Russians and being close to the USSR a lot of the things on sale came from there by truck across the border. The dress fabrics favoured by the women were in vivid colours and shot with gold and silver thread. Our friend Claudia just could not resist a pair of gold slippers on display.

Out in the desert, we encountered abandoned cities crumbling and vast. Wide streets between rows of destroyed buildings ended at gateways in the ramparts and old walls. The ancient wooden gates had gone but one could imagine travellers knocking on them to beg admission. Their populations had moved away centuries ago when rivers changed their ancient courses. Rivers with sources in the high snows of the Himalayas and the Karakorum plateau ran north to dissipate in the desolate sands of the Taklamakan desert. As the climate changed over the centuries or irrigation dams were built upstream, the rivers dried or meandered away, leaving the old oases waterless, and uprooting their populations.

From Kashgar, we were driven south and east over sand and gravel roads towards Khotan, where our guide insisted that we should enjoy the sight of a 400-year-old walnut tree. Our minibus came to an unexpected halt in the desert miles from anywhere where the flat grey wilderness stretched in every direction, absolutely featureless. The driver lifted the cover over the engine and pretended to search for a fault. In reality he well knew that we had run out of diesel but he wasn't going to admit to such rank incompetence. The men in our group managed to stop a passing truck and begged a can of fuel, the truck-driver agreed but insisted on collateral. We gaily offered them our long-winded guide as a hostage, until the next town. Thus, for an hour or two we did not need to listen to his monotonous monologues, so it was a perfect arrangement. By 'town' I mean a desolate windswept group of hovels cowering in the vast desert round a fuel-dump and a few petrol-pumps. A single track populated by starving dogs and ragged, filthy children disappeared in both directions into the howling waste of the Taklamakan. Taklamakan is well named: it means 'if you go in; you won't come out'.

In Asia, tourist-bus-drivers are a special breed, driving monoglot visitors across often appalling roads, sucking on their carious molars or cheap cigarettes, always uncommunicative and sullen. Unsurprisingly, but alarmingly, they often nod off at the wheel, so we always made sure that one of our group sat up front watching our driver. More than once Joan and I have been in a bus that's veered off the track and once we hit a street-lamp. In Mongolia a few years after this, Joan's tourist bus brought down the telephone wires and cut off communications for a hundred miles. They watched amazed as telegraph poles popped out of the ground, one-by-one, right along the valley ahead. These drivers didn't waste their time-off in sleeping and we often greeted ours drinking in a bar or disco in the wee small hours. We could sleep away the hours while on the road, no wonder the driver did too.

Eventually we arrived in Yarkand, an oasis surrounded by plantations of trees, where we stopped for lunch at the Mountaineering Institute. They fed us well in a canteen lined with faded and flyblown pictures of climbers including the great mountaineer Eric Shipton who had been a diplomat in Xinjiang in the 1940s. From the balcony we could see the snow-covered peaks of the vast mountain ranges in the south. The toilets were probably the most remarkable in all the world. Built outside on the lip of a precipice and protected from the wind in a sort of mat-shed, there were places for perhaps a dozen sitters, each separated by loose hessian curtains. It was built with long straight timbers, half tree-trunks, which were well-polished by use. In the gap between the slats, you could look down to where the ground fell away for at least a hundred feet. Below stood great black seething pyramids that must have taken years to build up. Their blackness was revealed to be due to a mass of flies which would rise up greedily when each sitter added his contribution to the pile, before settling back down again in a buzzing mass.

The next town was Khotan, where all I can recall was the hotel: vast, new and filthy; the vaunted river of jade-rich boulders and the famous walnut tree were certainly not worth a two-day bus ride across a featureless desert.

Returning towards home, we rode a steam-train for two days across the Gobi Desert, stopping to explore the western end of the Great Wall where, in the old days, China's territory ended. West of that imposing gateway was a barbarian and unknown land of danger and hostility. Exiles from the Middle Kingdom, whether criminals or perhaps disgraced administrators loaded with chains, had been expelled from civilisation through that gate. Otherwise, only a few intrepid traders with their camel trains would ever pass through to travel the great Silk Road carrying exotic silks and teas to sell to long-nosed foreigners more than a thousand miles away beyond the wastes of Xinjiang. Standing alone in the desert, the elaborate gate, newly restored and painted in bright colours for the tourists to admire, still felt a sad place. I could imagine a disgraced mandarin standing there in the sand, surrounded by his family as he faced the empty landscape that was to be his eternal punishment for failure.

On the train, we were glad to be in soft-class carriages with comfortable bunks and curtains to pull down for privacy. The railway company thoughtfully played music all day and all night through the P.A. system until I managed to disconnect the wires to the loudspeaker in our compartment. The local family in the next cabin enjoyed the music so much that they played extra music simultaneously on their cassettes. My rendition of the Goons' *Ying Tong Song* in protest at the din, eventually made them close their compartment doors and then we could sleep in peace.

Nazneen, our leader on the trip, had an endless supply of snacks that she handed out to us like an indulgent mother with an unruly brood of children. The dining-car was a few carriages ahead of us as the steam-train puffed and rattled across the Gobi. To reach it we had to navigate several cars of 'hard' class packed with travellers. They seemed to have a large supply of melons, no doubt from Hami, and the floor was knee-deep in their skins and sticky seeds as well as the usual detritus of paper, cigarette butts and things not to be looked at too closely. Before the 2-day journey was up, we declared the hard-class carriages impassable and waited until the train made one of its unexplained halts, when we climbed down onto the tracks and walked to the dining-car, worrying as we scrambled along about what would happen if the train re-started before we had clambered back on board.

The following year, Joan and I with another intrepid couple, returned to Kashgar from where we drove on to Pakistan over one of the world's highest mountain-passes and along the Friendship Highway through the Karakorum plateau and down into the Hunza River Valley through Gilgit. Historically, for the British, this was the North-West Frontier and the Hindu-Kush where the Red-Coats tried to keep the peace amongst the bandits and Pathan tribesmen along the frontier with Afghanistan. I don't think that even serious mountaineers visit the area today, and even in 1989 our visit was perhaps foolhardiness.

Pakistan, Joan on a typical suspension bridge over the Swat River.

We made many other trips in China, including visits to the south of the country where we visited Guizhou and Yunnan. The local people wore traditional dress even when working in the fields, the women in elaborate head-dresses and heavy silver bangles and jaunty hair decorations.

Near Guilin, we stayed on a hilltop looking down on rice terraces and then took the ferry along the Lee River where the fishermen used tame cormorants to catch their fish. At other times, we took less hardcore expedition routes and visited the fleshpots of Shanghai and Beijing with side trips to the gardens of Suzhou and to the Grand Canal.

In those years, before the millennium, and before the great expansion of China's economy, the tourist spots were generally quiet and unspoilt in contrast to the present influx of the masses. Nowadays, busloads of visitors, mainly local tourists, overcrowd everything. The old China seems to have evaporated but we were lucky to see just a bit of it before the 21st century swept much of it away in a tsunami of often vulgar affluence.

Chapter 28. Jake Joins The Practice.

We had a fine and varied group of doctors; the charismatic Anne in Repulse Bay, Fiona in Clearwater Bay, Tampa, Chris doing obstetrics and Bill still working hard, plus myself. We all worked well together and made a good team. Chris and I started to consider how we might replace Bill when he finally left us. We put out feelers and considered several candidates before we met Jake who was the senior surgeon at the local Military Hospital. Jake was due to leave the army after a distinguished career where he had been awarded the OBE for his work in Belfast, improving the immediate care of soldiers and civilians with gun-shot wounds. He was keen to remain and work in Hong Kong. Although there was no doubt about his surgical ability, both Chris and I had our reservations about his manner, but he was far away the best choice we had encountered so far and we were enthusiastic about recruiting such an eminent and broadly experienced surgeon.

Jake O'Donovan, a fine surgeon.

Bill retired to New Zealand where, with his usual assured energy, he plunged into the horse-racing world. Jake moved into the practice where he did not make the difficult transition from army officer to private doctor any easier by ordering everyone around. I spent long hours helping him settle into his new role as a civilian but he did upset some of the patients by treating them like squaddies and telling them quite abruptly that doctor knew best and to do what they were told. There was no doubting his clinical ability and he was an innovative surgeon who made his patients better quickly. Minimally invasive surgery[47] was now coming into the mainstream of practice and Jake was well placed to take this up as he was a skilled endoscopist with several years of experience. The medical school at Hong Kong U. offered training with practical instruction operating on live pigs[48]. After successfully completing the course Jake started removing diseased gall-bladders using his new skill and several surgeons referred patients to him while they awaited their turn to learn the new method.

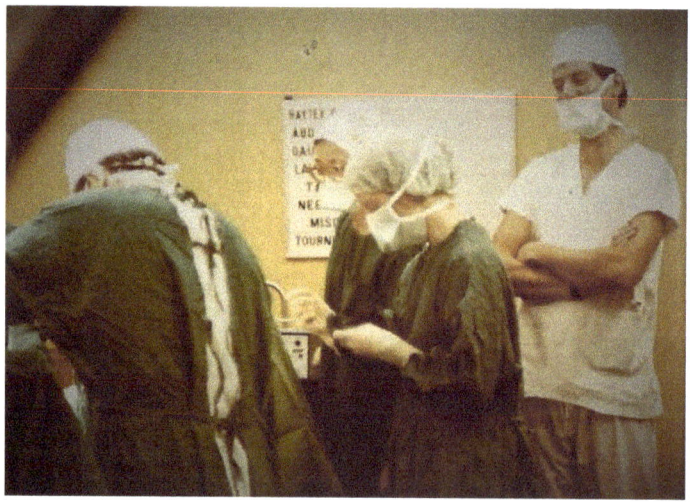

Matilda Hospital, Waiting to 'catch' the baby during a C-section.

Nevertheless, the happy team began to break up, starting with Chris who decided to leave and open his own office in nearby Duddel Street. This was a very sad event. I considered leaving too and joining Chris but felt I

---

[47] 'Keyhole surgery'
[48] Anaesthetised of course. These pigs, saved from the abattoir, lived long lives despite losing their gall-bladders in the name of medical science.

had too much responsibility for all our staff and patients to cause so much disruption and so decided to carry on.

We advertised for an obstetrician to fill the gap caused by Chris' defection and from a number of good applicants we appointed John. He was a graduate of a famous university in India, was well qualified and had also worked in Australia and in England. After leaving him for a few days to sort out his paper-work and local registration, I introduced him at the hospitals and to colleagues and so he began work. We had some initial friction when he wanted us to employ his wife to be his nurse but that was soon settled and he became busy. John was popular with everyone and he attracted a new group of patients from the local Indian community.

Then I started having doubts about him. The practice had a rule that upon the death of any patient, the event must be reported to all the other doctors for the simple reason that we all needed to know in case we were consulted by one of the deceased's family who might, quite rightly, be upset if we seemed unaware of their bereavement. When one of his babies was stillborn and I only learned of this second hand, I was furious and remonstrated with John, who seemed contrite.

A year or so after this event, a colleague from another practice tipped me off about a major row brewing over John whom he claimed was practising unregistered. I checked with the registration office and indeed, John had applied for registration but had failed to follow it up fully and so was still effectively unregistered. In the eyes of the law he was practising as an 'illegal midwife', a criminal offence. At our inevitable interview, John told me of his day applying for registration and filling out forms, claiming he had assumed that that was sufficient and had not bothered completing the process which was more onerous than it had been for me in 1974. I had no choice but to fire him on the spot and he promptly left for India. Unfortunately, it was not to be the end of the matter.

About twelve months later, an official letter came from the Medical Council Disciplinary Committee, the letter that every doctor fears to receive. Jake and I were summoned to explain why we had employed an unregistered doctor in the practice. The committee had received a complaint that had to be followed up. Our defence society appointed a barrister to advise and represent us and we tried to defend ourselves as being in fact among the victims in the matter. John was undoubtably a well-qualified and skilled surgeon, but we had to admit that we had failed to check on his registration before allowing him to treat patients. We were duly admonished and the hardly sensational story made page 3 in the SCMP. I did not expect to hear anything further about the matter, putting it down to experience. However, I was still interested to know who had made the complaint, although the complainant is allowed to remain anonymous. In the aftermath, I asked my colleague at the hospital if he had had anything to do with it, but he denied that either he or his group were involved, although they would have indeed considered reporting us if we had been slow to take action. Later, while tidying up the voluminous paperwork from the Medical Council, I came across a Council memo that should not have been given to our lawyers. It named the complainant as John himself.

My reaction to this was to inform the three doctors who had provided John's references when he had applied for the job, informing them of his duplicity. In the meantime, Keith, an obstetrician from Belfast had taken over the gynaecology and obstetrics and our practice ran smoothly again treating the people who came into our clinics, day in, day out.

If you were a cynic, you might think that hypochondriacs would be the most welcome patients a private doctor could have. I can assure you that in my case this is not true at all. Salvatore Ferrari arrived in Hong Kong with his small son and delightful Korean wife. She had been a fashion model in the top flight of the profession, appearing in Vogue and on the catwalks all over the world prior to settling down as Mrs Ferrari. Salvatore was a trade commissioner at one of the consulates and was always dressed immaculately. A big man, blonde and looking very fit, his long hair drooped across his brow, his suits fitted perfectly, and he usually sported a silk cravat at his collar. He had tiny feet for a man his size and these were clad in the lightest Italian leather loafers. He must have spent more time in my clinic that he did at the Consulate and he drove me crazy with his symptoms and those of their little boy. One of his obsessions was with bodily waste or 'fees-seize' as he called it. He had installed special toilet bowl in his apartment, with a flat shelf strategically placed to catch the 'fees-seize' and allow him to inspect not only his own but that of his wife and the child. He would then phone me with a daily update. One day he was worried that his own productions were 'hollow'. I made the mistake of asking how he could possibly know this and he embarked on a long description of his methods of dissection using Japanese-style chopsticks. This all ended with his sudden disappearance from Hong Kong when he found his wife was having an affair with a fellow Korean and was buying him expensive clothes on their credit card.

Salvatore was not my only trial. Savita, a dumpy Indian lady, was bringing her little boy to see me for various health issues. She herself had trained as a doctor in India but had never practised because her family had arranged a marriage before she became too independent. Clearly, she regretted the way her life had gone and she would describe her husband as a pig with poor hygiene and other bad habits. I never met him and had to take her word for it. She then declared her love for me and started to write me love-letters. She would phone Joan at home demanding to speak to me, informing her that she was coming to live with us and had her suitcase packed. The situation came to a head when she had to be removed from our waiting room by a security guard. She was lying prostrate on the waiting room carpet, gripping my ankles and wailing her love at full volume. I had to organise a restraint order from a magistrate in the end to keep her away from my clinic. At least it kept the waiting room lively.

Young ladies could be a nuisance but old ones also sometimes caused problems. Lady Williams was the widow of a retired colonial officer and was living out her life in a tiny flat in Repulse Bay, looked after by a devoted Chinese amah only slightly younger than herself. All her family had long before disappeared to London. I visited her from time to time at home to update prescriptions and keep an eye on her ulcerated leg. At each visit she 'had something special for you, doctor'. A jewellery case was opened and a glass bauble displayed and handed over against my protests. None of these items was of any aesthetic or monetary value and I put them in envelopes with the date and my nurse would countersign them. This went on for a year or two until she died and my desk drawers held a fine collection of Woolworth's best costume jewellery. I then received a solicitor's letter from London accusing me of taking advantage of a dotty old woman and stealing the family's inheritance. It was all sorted out in the end and the solicitor apologised, but the family did not. I think they still considered me an evil Svengali. As is so often the case the family had ignored the old lady, leaving her to the care of the amah and myself while they anticipated the wealth they were to inherit.

The girls were growing up and in 1989 Diane was ready to leave for university in Canada and was taking a gap-year as young people do. She was having a great time and had a job waitressing in a bar in Wanchai while she saved up to go travelling. Julia was near the end of her time at Bradbury Primary School and at the end of May, flew to Beijing with the rest of her class to see the sights. They visited the Temple of Peace, the Great Wall and the Forbidden City next to Tiananmen Square.

CHAPTER 29. A NEW BOAT AND NEW PETS.

Despite all the difficulties in the practice, our family was getting along well. Joan's business was thriving and taking on an increasingly large staff. She moved offices as more space was needed, bought herself a better car to transport her clients around and then became involved in fund-raising for a charity. Her great friend Vanda had risen through the ranks of Samaritans, the suicide prevention charity, and was now a leading light in Befrienders, an associated international group combatting suicide round the world.

Joan and Vanda leant on their contacts to sponsor a great event, a May Ball, to raise money for Befrienders. The army offered the parade-ground at Stanley Fort. It was perfect: quiet, private and commanding a wonderful view overlooking the sea and beyond to the lights on the Peak. Caterers were hired, booze was donated by importers, bands were recruited and we prayed for good weather. The Befrienders May Ball unfolded in style and the glitterati of Hong Kong society ate, drank and danced the night away under a sky filled with twinkling stars. A large sum was raised at an auction. Everyone was happy, we had a wonderful night with dinner and music, and a stack of money was handed over to the Befrienders afterwards.

Government House, Lavender & Chris Patten with
Joan on the right.

Christmas cards were the next money raising scheme. A local artist provided the designs of Hong Kong scenes with amusing pictures of Santa's sleigh landing near a local landmark, all in bold colours. They were to be sold at St. John's fair, a pre-Christmas event at the Anglican cathedral. The printer promised delivery well in advance to give us time to package the cards and envelopes. Then came the delays: the fair was on Saturday morning and on Friday night the printer was still promising delivery. Joan was tearing out her hair in frustration, almost weeping. At about 1a.m. the printer's van arrived and the whole family spent the rest of the night putting cards and envelopes in their packets. They were beautifully produced and at the fair, the Befrienders' cards outsold the other charities. It was another success for Joan who continued raising funds for Befrienders until we left the Colony, including a gala dinner at Government House hosted by the Governor and his wife, Chris and Lavender Patton.

Then I heard on the grapevine that *Sparrow* was for sale. She was a 38-foot cruiser/racing yacht with a successful record in local races. I went to see and took her out for a sail. Built in Taiwan to a New Zealand design she was a tough boat and seemed a good choice for my first large yacht. I moored her at the Aberdeen Marina[49] and we raced with the Aberdeen Boat Club in their Sunday events.

Zest with her racing crew.

---

[49] Aberdeen, the fishing harbour in Hong Kong, not Scotland!

I gathered an enthusiastic crew and we started to do quite well once we understood how to make her go properly. Often, I was delayed at one of the hospitals on race mornings and so *Sparrow* would sail without me as I toiled in the wards or delivery room. This suited my crewmembers who got to take command. One of the crew worked for an American brewery and every week brought a case of Budweiser. Since everyone preferred Carlsberg, the Budweiser accumulated in the lockers where its weight must have slowed the boat's performance quite a bit.

One long holiday weekend the family sailed the 40 miles or so to Macau where the harbourmaster greeted us over the VHF radio calling us 'Spah-loh, Spah-loh, Spah-loh' several times as we approached, giving us navigational advice and guiding us to our anchorage. I hadn't liked the boat's name in the first place but this rendition was too much and soon we painted her new name on the transom: *Zest.*

Zest sailed several times to the Philippines and back and I learned a lot from sailing her. Certainly, I made many mistakes but she carried us safely for several thousand miles without serious mishap including sailing across the South China Sea during a typhoon (another error of judgement!)

Spitz was 14 and growing old, so we thought a second dog would be a good idea. Perhaps Spitz would help educate it with his better habits, though I actually doubted this since he had so few. Tosh joined the family; a little black Chinese chow-dog with a black mouth and a tightly curled tail. She was from the litter of Blackie, the guard-dog at the Middle Island base of the Yacht Club. Julia was having sailing lessons on the island and decided that Tosh was the best of the litter. Tosh moved into the Sassoon Road house and was accepted by Spitz and Middy once they had shown her who was to be boss.

There was one problem with the house, the basement area was not originally intended as a kitchen and had not been provided with proper rainwater drainage. This was only a nuisance in typhoons when heavy rain leaked in and the kitchen floor ran a couple of inches deep in water with Spitz in his basket floating around looking miserable. We never found a solution but it only happened a couple of days a year so we put up with it.

Julia liked to bring home stray dogs and cats. Usually, they were fed a few times then encouraged to leave, but a big mongrel she christened Charlie stayed with us for a year or so before meeting his nemesis in the form of a speeding truck. One morning he was in the back of the car on the way for a walk on the Peak where I paused briefly at the Matilda to check on a patient and left the car next to the railings where the carpark was built up about fifty feet above a steep cliff. A mature tree, its roots far below on the hillside shaded that corner but Charlie did not notice the drop before he jumped from the car window and over the parapet. He crashed down through the branches of the tree and then rolled down the slope below. When I returned from seeing my patient, he was still whining far below us and I had to clamber down and rescue him. He lay there shivering and looking pathetic but refusing to help himself so I had to carry him up the snake infested slope, exhausted from the climb with my clothes filthy.

At home, we employed an excellent Indian lady as a helper, her rather pathetic son in his late teens shared her accommodation and we paid him an allowance for doing small jobs including exercising Charlie. One weekend I was enjoying a late breakfast in the kitchen, when a police sergeant knocked on the back-door and told us that he had come to arrest this boy. Apparently, a Filipina woman had complained that he had indecently assaulted her. She had been standing at the nearby bus-stop when, she said, the boy touched her on the bottom, right between her legs. 'No,' the policeman insisted, we could not speak to her and the boy had to go straight to the police station to be formally charged. The boy returned later in the day after had I arranged for a solicitor to assist him. In due course he was summoned to appear in the magistrate's court. We could not believe that this very weedy boy would have the gumption to goose a mature woman's bottom. We suspected that in reality it had been that damned dog, Charlie. One of his habits was indeed, poking his nose into unsuspecting people's bottoms. Anyway, although he had good legal support, the boy took fright and insisted on returning to his father back in India and so he skipped bail. His mother, mortified by all this, left us soon after. I was very sad, especially as she was a wizard Indian cook and her stuffed chapattis were to die for.

Paediatrician at work.

Although he had left the practice, I still worked a lot with Chris in the delivery rooms as well as with other doctors and also John's replacement in our practice, an Irishman called Keith. For obstructed labours, I was accustomed to Chris' practice of using special straight forceps to rotate the baby. Keith preferred using a Ventouse, a device with a cup that latched onto the unborn baby's scalp by suction. Few doctors today are trained how to use straight rotation forceps and more and more they turn to Caesarean deliveries. I think this has meant that in the area of obstructed labour, many old skills have been lost and, in another generation, hardly an obstetrician will know how to use forceps skilfully. The change has been driven by legal constraints of course. It almost seems that when there is a damaged baby and that if counsel mentions forceps, the case is lost and huge damages demanded.

I have worked with many obstetricians over the years and have observed much variation in their abilities. Some seem never to have any trouble and they foresee problems before they develop, while others always seem to be catching up, with disaster already upon the patient before they have seen it coming. Unfortunately, some of the most fashionable doctors with the smoothest manner were in the latter category.

Doctors since time immemorial have been subject to the attentions of lawyers acting for patients with grievances real or imagined. Of course, if a doctor does cause harm to his patient, due to incompetence or negligence, then there must be due apologies and recompense for the damage, and that is why we have insurance through medical defence societies and unions. Most cases can be settled privately but many do have to be decided in court. I am fortunate that in my years of practice I was never sued.

A large law firm would from time to time ask me to review paediatric cases they were handling, usually on behalf of the plaintiff, a patient or more usually a parent of a patient whose care had not gone well. A mountain of paperwork would be delivered to my office: medical notes, nursing notes, letters, copies of legal documents, prescriptions, invoices, and lab. results. Usually in no particular order and often running over a long time period. It all had to be sorted out so a clear picture of events could be understood as they had happened. I needed to appreciate events as the doctors and others perceived them as they occurred at the time without the advantage of hindsight. It is an old joke that the 'retrospectoscope' is the most useful diagnostic device in any hospital. Anyone can solve a problem when they already know the answer.

It is true that sometimes patients do suffer unnecessarily and doctors are culpable, in which case it is best to settle these cases preferably out of court as quickly as possible so everyone can get on with their lives. However, sometimes the wrong doctor is being blamed. Patients seemed prepared to put up with all sorts of abuse from a charismatic doctor whereas a mild self-effacing one can find himself getting the blame for the mistakes of a more outgoing doctor.

In one case I had to review, the obstetrician had behaved disgracefully and the baby was near death when eventually delivered. Despite the excellent care provided by the on-call duty-paediatrician, the baby boy was severely brain damaged. The parents then sued the paediatrician and would not accept that 'their' doctor, the obstetrician, had been totally responsible for the disaster. Their case had no chance of success and was abandoned, although, had they taken proper advice, they would have gained huge damages for their son.

One of my patients, a solicitor, begged me to take an informal look at the case notes of a well-known Chinese tycoon who had recently died after a long stay in hospital. She had a serious problem on her hands, not of medical negligence but of determining the state of mind of this great man. She had shown the notes to the doctor who normally advised her firm but he had been unable to help on this point. It seemed that the dying man's hospital room had been a magnet for employees, business associates, relatives, ex-concubines and multiple illegitimate children. They had approached his bedside by turns with their supplications: disbursals of money, property or valuables were added to his will every day. His secretary had sat there in the room recording it all. Names, dates, legacies were all meticulously listed. The problem was that the book entries continued over many weeks and one of the deceased's immediate family, who had been watching the old man give away fortunes, now insisted that for a lot of the time he had been comatose and quite unaware of the proceedings. This challenge was causing great difficulties and was heading for the courts. I asked to be shown the medical notes and assured them that we would soon sort it out. How wrong I was! The medical notes, which covered the almost twelve months that he had spent in a hospital bed, might as well have been blank sheets of paper for all they were worth. Not a day had passed without at least one doctor adding his name and a figure of so many dollars. That was all: a note of the fee but not a single clinical observation or order, around fifteen different doctors all charging their daily fee. The nursing notes, all in English, consisted solely of his fluid intake and urinary output. However, I noted that about two months before he died the nurses had stopped measuring his urine volume and started weighing his diapers. Although nobody had bothered to record his mental state, they still wrote down his urinary habit. In the end the court accepted my suggestion that they used the 'nappy day' as the cut-off point after which the gifts were void. In this undignified manner, a great tai-pan died.[50]

---

[50] His carers even got the diagnosis wrong, he died from an amoebic abscess of his liver, not cancer.

One season followed another; year followed year. The girls grew and developed their own personalities. Our family were not free of the usual adolescent problems and Joan and I worked hard to keep us all happily together. We decided that holidays must be fun for all of us. The girls did not really like sailing, though using *Zest* as transport for swimming parties to outlying islands was very popular. We took up skiing and bought a tiny condo in the Canadian ski-resort at Whistler Mountain. Truly the family that played together did stay together and the girls liked to holiday with us during their university years and later on boy-friends and husbands came too.

Zest sailing through a choppy sea

One day in high summer, just before the schools broke up for the holidays, a Russian cruise-ship moored in the Harbour with dozens of school-children from Vladivostok aboard. They were under the impression that they had invitations from some Hong Kong schools to come and visit for a week. The Immigration Department could find no evidence of this and proposed to keep all the kids, (about fifty of them), on board until the ship was due to sail back home. The headmaster at Island School then intervened and arranged for them to come ashore to stay with pupils and staff from his school. Sascha, aged 17, undertook to host Yelena for a few days chez-nous.

Yelena was a big girl, dark complexioned with a faint moustache and hairy legs. She spoke excellent English and was a cheerful girl. One morning I went to do the family shopping at Park N Shop, the nearest supermarket in nearby Baguio Villas, and took Yelena along. She was looking askance at some of the other customers who were in their rather scruffy working clothes only to tell me loudly that such persons would certainly not be allowed in the shops where her family and other party members went. I replied that I was pleased to hear that the fall of the communist state in Russia had not made anyone less equal than they had been under the Soviet system. She probably hadn't read any George Orwell, so I suppose such irony was lost on her.

B.O. was another of Yelena's problems; not to her but to the rest of us. She stank. Sascha and her best friend, Nina, tried to subtly suggest she make more use of the en-suite shower in her room and lent her their deodorant, which she immediately scorned. That afternoon, out on *Zest,* we all kept to windward of her as best we could while Emily and Sascha tried to persuade her to have a swim. This was despite the reports of recent shark-attacks on swimmers at Hong Kong beaches: it was her or us.

I generally avoided looking after eye problems so usually passed such patients on to an ophthalmologist. Inevitably patients with minor eye injuries or infections came to us and I would deal with those but proper diagnosis really necessitated using special equipment that ordinary doctors like myself were not trained to use. Michel, a French Canadian, had persistent conjunctivitis that would not settle. Charles Chow, my go-to eye-doctor, recommended a course of Cromoglycate eye-drops and these seemed to work as long as Michel used them regularly, so he came in occasionally for a refill of his prescription. One Monday morning he appeared looking terrible. He was unshaven, and his eyes were red and swollen and he had quite a tale to tell. He had been alone in his flat on Friday night as his wife and daughter were away on a school camping trip. In the night his eyes became a bit itchy so he felt around in the bathroom cupboard in the dark for the little squeezy bottle of eye-drops. Here he made a terrible mistake; the plastic bottle he picked up was actually of Super-Glue. He squeezed a few drops into each eye and went back to bed thinking it stung more than usual. He understood what he had done when he realised that his eyelids were stuck closed. He spent a while trying to wash away the adhesive and then decided to phone our practice night-call number. Blind as he was, he realised the only number available to him was 999. When the ambulance came it took him to the government hospital near Happy Valley. The nurses were very kind to him. They bandaged up his eyes, took away his clothes and dressed him in hospital pyjamas, inevitably, a few sizes too small for him.

Still unable to see, he could not get anyone in the ward to operate the phone for him and over the weekend there was no doctor to speak to. He knew that his family would be back home on Sunday and would wonder where he was. Indeed, his wife, after finding neither note nor husband, did ring round his friends and then the private hospitals and finally the police. Michel, getting frantic, decided he must escape, but his clothes, his house keys and wallet were all locked away until Monday. Desperate by Sunday evening, he peeped through his bandages and found he could see! He was out of bed in a moment, down the stairs and flagging down a taxi in the street outside. He arrived home looking like an escaped lunatic, dishevelled in unbuttoned nightwear and no shoes, his unruly hair anyhow and wild red eyes. Michel's wife went down to pay the taxi but it had already fled; the poor driver must have been terrified by the wild gweilo he had picked up.

Britain and China were into the final negotiations for the handover, 4 or 5 years away. China of course held all the best cards and behaved accordingly, while Britain's team tried to play from the moral high ground; but attempting to promote democracy for Hong Kong sounded a bit hollow when one considered the lack of progress in the local electoral process even after 150 years of colonial rule. Chris Patten, the governor, had opprobrium heaped upon him by the Chinese; a high point of which was when he was called 'a whore for a thousand generations' during this phase of megaphone diplomacy. We citizens could only try to be optimistic as negotiations proceeded filling the pages of the SCMP with memorable quotes and vituperation. Mr. Patten's woes were not eased by various senior Foreign Office diplomats, notably Sir Percy Cradock, who were happily briefing against him and, probably concerned about trade balances, seemed to have the interests of Beijing at heart rather than securing a fair deal for the people whom they were intending to hand over. In the end the 'One Country Two Systems' slogan was hit upon to last for 50 years, but without any convincing safeguards to ensure it would be upheld.

One evening during a particularly stressful phase of the negotiations, Chris Patten with his wife Lavender, came into a restaurant where Joan and I were dining with friends. The entire clientele and also the waiting staff stood and applauded until he had taken his place. Everyone in Hong Kong had great respect for his effort even if some of us had doubts about what his brave resistance and his dogged insistence on the rights of the people of the Territory might bring about.

'A tiny kitten clinging desperately': Tenzing the cat.

Hong Kong continued to prosper despite the dramatic diplomacy swirling around us all. Joan's business, Relocations, had opened offices in Shanghai and Beijing and she was proposing branches in Indonesia, Malaysia and Singapore too. I was as usual working as many hours as the

Lord provided and we were all happy enough, especially living in the leafy and quiet area around Sassoon Road. One evening in the aftermath of a typhoon, Joan and I were having a walk not far from our house and enjoying the few hours of cool air before the sun broke through again. Rainwater was still swirling along the gutters and roaring through storm drains but it had stopped falling when I heard a pathetic mewing cry from somewhere in the undergrowth. Exploring, I found a tiny kitten clutching desperately at overhanging vegetation, its body was half immersed in the rushing water of a large drain. Another few minutes and it would have been lost to a watery death. Rescued, it nestled in my hand, shivering but watching me alertly. I wrapped it in my shirt and took it home where it lapped up some milk and fell asleep in a cardboard box in the garage. The next day, there he was waiting for breakfast and after we had given him a thorough bath and search for fleas, he took over the kitchen. I don't know how he did it but the dogs now had a new boss. We gave him a dish of his own but he preferred to use it as a bed in which he could curl up; he continued to use this bowl to sleep in as he grew to an adult but always lay across it somehow, looking very uncomfortable. Cat-food he refused, preferring to use the dogs' bowls as his own, not allowing them to start their meals until he had finished. The girls named him Tenzing as to their eyes he looked Tibetan. The dogs didn't seem to mind his dominance especially as he had, soon after arrival, discovered how the backdoor-handle worked, a feat they had not mastered in many years in the house. They could now wander in and out at will once Tenzing had performed his trick for them. The cat himself preferred to use the kitchen window as an exit. He lived with us for several years and then suddenly disappeared. Our security guard suggested he had been attacked and killed by a pack of feral dogs that roamed nearby. We missed him enormously.

CHAPTER 30. CODA.

Time was passing and it was only two years now before the 'Handover'. It was on everyone's mind and lips. My old friend Chris decided that he would not work in Hong Kong any more after the British left and others voiced similar views. My attitude was different: the political aspect did not seem to be the problem at all as the administration was supposed to continue unchanged for many years despite a red flag flying where the Union Flag once hung.

Was this truly how Julia saw her dad on his 49th birthday?

Our daughters were either leaving for universities abroad or had left already. Diane was in London at LSE studying for her master's degree in Economics. Julia the youngest was due to start medical school in Dublin in 1998 and was going to Chile on an expedition in 1997 after A-levels. Sascha was at Art School in England where later she became a jewellery designer with her own small business. 15 years later when she and her family returned to Hong Kong for a few years, she opened her own art studio in Sai Kung and became well known for her ink-wash paintings of the local cows that wandered free in the area.

I was finding it increasingly difficult to work with my partner Jake and a lot of the joy in our practice was, for me, fading away, week by week.

From a clinical point of view, we got on well with mutual respect but on a personal level, things were going less well.

We were seeing a lot of patients; our bookings were full and our financial situation was good, but it wasn't half the fun of earlier times. 12 doctors, some of them part-time, plus about 50 other professionals and support staff were on the payroll. We had moved into a larger space in Bank of America Tower and a young Australian cardiologist, another Chris, set up a small heart lab, including a treadmill for stress testing.

Personally, I was still at full stretch as usual. None of these new doctors seemed to reduce my clinical case load. They were working hard too but my own following stuck with me. This was flattering indeed and I was always delighted to see my old patients and also to greet new faces. But when a long-standing patient died, it was a wake-up call. She had been a patient of the practice since before my arrival in 1974 living across the road from our clinic in Repulse Bay. She suffered from an inherited disorder that caused slowly progressive lung damage. Over the years she had steadily become more breathless and in truth had outlived her expected span by some years despite being a heavy smoker. One weekend, when I was absent, she had been admitted to hospital with a chest infection by one of our doctors and despite good management died within a few hours. My presence could not have made any difference to her chances of recovery but the fact was that I felt that I had failed someone who was an old friend as well as a patient. The practice was just getting too big and becoming impersonal.

A lot of my work was still out of hours in the maternity wards. Babies arrived at their own pleasure and about twenty percent of them needed a paediatrician standing by because of potential problems during their delivery. Our obstetricians had low rates of operative deliveries but I was still out of my bed several nights a week and I began to feel the strain. After all, I was over fifty years old by this time and less able to recover from sleepless nights followed by full days in the clinic. My eyesight was a new problem as inevitably senile presbyopia began to interfere with my near-vision; spectacles worn above a face-mask steamed up at critical moments as I inserted endotracheal tubes. It was getting harder to see babies' vocal cords. I didn't feel old but time and tide………

We were then swindled in a small way by an office manager whom I had trusted as a personal friend and I began to wonder if I was no longer up to the pressures of the job. Several nearby practices like ours had been taken over by insurance companies and investors and some had started a pricing war amongst themselves. Initially this did not affect us, though I could foresee a time when it might. Jake began to talk of selling the practice for 'big bucks', but unfortunately he did not do this privately so when the staff heard of his ideas, several key people decided to resign. Patients also spoke worriedly about rumours they had picked up on.

Then Anne came to me, she was absolutely furious with Jake and had tendered her resignation. In fact, she just walked out as she was so very upset. Her version of the exchange with Jake began when she referred a patient with appendicitis to him for a surgical opinion. Instead of being pleased about her confidence in his skill, he ticked her off for saying the man was 'her patient'. Jake admonished her telling her she was just an employee and the patient was not hers but belonged to the practice. While this was all technically true it was an insult to an able, committed and loyal colleague and should never have been said. Inevitably within a few weeks Anne put up her sign in a nearby building and many of her patients followed her. I could not blame them, nor Anne either. Once again, I felt that I was losing control of the practice, that Jake was a loose cannon and that it was time that I should hand over the practice, or at least take a sabbatical somehow. I was very conscious at the time that a number of sailing buddies of about my own age had fallen seriously ill and two or three had died in the last few years. Perhaps this was now the time to fulfil a long-held ambition to go sailing for a year or so. I was already thinking of selling *Zest* and ordering a larger cruising yacht suitable for blue-water sailing from a builder in New England. The late 1990s seemed to be a good time to at least start a circumnavigation if ever I was going to do it.

The new boat, '*Ming,*' was launched in the spring of 1996 at Ted Hood's shipyard in Plymouth, Massachusetts. Joan and I intended to spend our two weeks summer leave on board getting to know the boat and taking her for a shakedown cruise up to Maine. Our cruising plans were frustrated by a number of breakdowns and problems that perhaps we should have anticipated in a new design of yacht. We did manage to sail round Martha's Vineyard and Cape Cod between visits back to the boatyard for necessary repairs and modifications. The thick sea-fogs, which are common in the summer there, provided useful learning experiences as using radar and GPS, we felt our way from port to port and from buoy to buoy.

I hoped to start our cruise in the spring of the following year, visiting the East coast of America and sailing down the Intracoastal Seaway and through the Caribbean, eventually sailing west into the Pacific via the Panama Canal. However with the delays in arranging the future of the practice I had to change my plans and *Ming* was ignominiously loaded onto a cargo ship bound for Hong Kong.

Arriving Hong Kong, Ming is swung into the water

One afternoon, just before Christmas, with the help of Yacht Club staff, I collected *Ming* at the Kwai Chung Container Port. She was rolled off the huge Ro-Ro ship accompanied by dozens of double-deckers on delivery to the local bus company. She was craned off her cradle and into the water and her mast was lashed down horizontally along the deck. We opened her

250

seacocks and the engine started with the first turn of the key. Accompanied by the Club's tender, *Ming* motored through Victoria Harbour to Causeway Bay where, in the Club's yard, the mast was stepped and her bottom was given a new coat of anti-fouling paint. Soon I was able to take her to her new mooring on the south side of Hong Kong in Tai Tam Bay where I could visit her frequently and make some necessary additions including air-conditioning and a 220-volt generator. At weekends we could sail round Hong Kong waters. With Joan away at work I often sailed single-handed and learned much about handling her, and was able modify her rig and her systems to suit my ideas and style.

Her name, '*Ming*,' in Chinese means light and brilliance; brilliance in all the ways we use it in English too. The ideogram incorporating those two great sources of light, the sun and the moon, is regarded as very lucky in Chinese tradition and, painted large on *Ming's* stern, drew approving looks from Chinese friends and Yacht Club staff. She was a 56 feet long Sundeer, a design from a maverick called Steve Dashew who championed yachts that are fast, safe and capable of being managed by a small crew. He and his wife Nancy, both a little older than Joan and myself, had sailed a larger prototype version of *Ming* many thousands of miles together. *Ming* was to prove herself the perfect cruiser and safely took us thousands of miles as well.

Aboard Ming, Julia 50 feet above deck

251

Meanwhile, our youngest daughter Julia, was training hard for her expedition to the mountains of southern Chile and doing a lot of hill-walking round Hong Kong, usually with a backpack weighted with telephone directories. One morning I joined her on one of our favourite hikes in the New Territories up Robin's Nest Hill near the closed border area. From the top of this beautiful and wild place, we looked down on the fences and sentry boxes that controlled the border with China. To the north-west was the new Chinese city of Shenzhen. Modern office blocks and apartments towered above a network of multilane motorways where only a few years earlier we had watched farmers with water-buffaloes tending fields of vegetables or rice. Further east, in Starling Inlet, was a brand-new container port with huge ships plying the channel. However, the view to the south, back into Hong Kong, still looked much as it did on our first visits to the New Territories more than twenty years earlier with sleepy old villages, narrow tree lined roads and here and there a farmer hand-watering his crops. What a contrast! China was now the moving place, the economic power. Hong Kong looked a bit drab and old-fashioned from up there on Robin's Nest. Wasn't it time for me to move on? I too had changed. I was beginning to lose my enthusiasm; I was no longer that young doctor brimming with new ideas who had arrived in 1974 with little else but a lot of hope and energy. Had I already given all I had?

I continued talking seriously to Jake about selling my share of the practice and handing everything over to him, but this did not seem to be his ambition at all. He certainly liked to be involved in decision making, but he did not want the responsibility and chore of running the whole business. After several weeks of talking round the subject and getting nowhere, I had pretty well given up. Jake was happy with the status-quo. He could see his patients, work in the operating theatre, order people around a bit and draw his monthly cheque. He wanted a say in things but no more than that. In retrospect I can see that understandably he wanted to continue as we had originally agreed and now saw no reason to change. In fact he considered that I was letting everyone down and was in effect deserting the practice. Perhaps he was right and I was being unreasonable but there

252

was no-one else to delegate to. Totally fed up with my partner, I was nearly in despair.

Then along came a large medical group from Singapore who wanted to expand their business by buying a practice in Hong Kong. After inviting us to Singapore for a tour round their superb facilities, they made an offer to take us under their wing. Their policies were unlike mine but I agreed to accept their terms although the process took longer than I had expected.

My last months in the practice were not happy ones. I had surrendered control to the new owners. I certainly have no criticism of them as I had knowingly sold out my staff and my patients to a new master who intended to do things differently and probably more effectively. However, I felt guilty for abandoning the nurses, technicians, doctors and office staff who had worked loyally and enthusiastically with me, some for many years, and had shared my vision of how medical practice should be. I could afford to leave if I wished and to move on, but the change was forced on them.

The patients were a concern too. Some of them I had known since the 1970s, some I had looked after as children and were now starting their own families. I felt that I was leaving them all to fend for themselves and to get used to a new doctor. In my need to leave I recognised that I had given little thought to those that I was leaving behind, but the decision was made.

I left the practice on the last day of June 1997; the day before Hong Kong came under Chinese control. The coincidence was just a coincidence; negotiations had taken months longer than I had anticipated.

That evening it poured with rain as it only can in the tropics. Joan and I had seats to enjoy one of the open-air handover ceremonies and we sat drenched to the skin despite our rainproofs and umbrellas. We watched schoolchildren dancing under the lights in the downpour, the bands played and soldiers marched. Chris Patten, the out-going governor, streaming wet in the rain, told us without irony that it was a 'happy day for Hong Kong which would now be ruled by Hong Kong people'. It felt like the end of Empire.

Ming at her mooring in Hong Kong.

Overcome by nostalgia, we ran back to my old office in Central to dry off and put on our party clothes for a dinner-dance at the Hong Kong Club where 'après-nous le deluge' was very much the general feeling. We danced the night away as Chris Patten and his family, reportedly in tears, drove from Government House to sail away with Prince Charles aboard the Royal Yacht at midnight. Hong Kong, the Pearl of the Orient, Palmerston's 'Barren Rock with Nary a House Upon it' was now, after a century and a half of British rule, a region of China with an uncertain future, dependant on an untried policy of 'One Country, Two Systems'.

Next morning the sun shone and as I enjoyed a lie-in, I realised that I was unemployed and without a job. No patients would be calling. The phone stayed silent. But my new boat lay at her mooring in Tai Tam Bay and I was as free as a bird.

EPILOGUE.

Six months later Joan and I sat in the shade of *Ming's* large comfortable cockpit. We were anchored off an island near Phuket. The sea was crystal clear and we could see, three or four metres down, the sandy bottom of the bay. We had just rowed back from lunch at the 'Same Same But Different' restaurant on the beach. Aboard *Ming* we had sailed away from Hong Kong in November 1997 with stops in Singapore and Malaysia. In a few days' time we left Thailand and crossed the Andaman Sea bound for the Maldives on our way to the Red Sea, the Suez Canal and the Med. Our life in the Far East was coming to a close. Both our businesses had been sold to their new owners. Our children were scattered across the world, either studying or working. Even the little black dog Tosh had relocated to England and was living with Joan's parents in Liverpool.

But that's all another story. After three years of cruising aboard *Ming* we eventually reached the Caribbean. In a later story we will come back to Hong Kong for another two happy decades, no longer just parents but grandparents.

THE END.

Author's Note.

I would that I had written this note before starting out to write my book, for then I might have known what I was trying to achieve. It's not a novel, but perhaps a social history, or a memoir? Is it comic, or perhaps Comedie Humaine?

First of all is it accurate? No; but then it cannot be. Just the necessity of providing anonymity, even past the grave, for many of the protagonists, be they patients, medical men and women or innocent bystanders, renders strict accuracy an impossibility. I have had to change not just names and nationalities but the chronology too. Even the topography has been manipulated. If time and place are altered then what hope is there for truth and accuracy? And then there is my memory!

What I have tried to make true is the feel of the times, how medicine was practised, at least by me at the time, interlaced with my thoughts, my motivation and even the morality of thought and action, both mine and others.

Everything described in this book did happen but, as I've noted above, anonymisation has changed details but not what really was going on.

If perchance you think you have identified yourself or another in the events described you are certainly wrong. These tales are combinations of multiple events for, to be sure, in medicine as in life, there is nothing new under the sun. It's called experience and that's how doctors know what they are doing (most of the time).

I have identifiably included some real people in the book, amongst them my family, friends and doctors with whom I was associated and whom I admired. Other medics are renamed and I hope rendered anonymous in other ways, though they may themselves recognise events in which they participated. Hospitals are named and are described as I remember them at the time; over the past two decades standards have changed so much to the extent that all hospitals and the care they provide are not comparable with the days I describe.

I hope I have represented real people on the world stage and in Hong Kong and elsewhere correctly and I apologise for any inaccuracies I have included.

Hong Kong and medicine have changed very much in the two decades since I retired and part of my aim is to tell the story of medicine and of Hong Kong as they were between the 1970s and the millennium.

There are a few people I must mention and thank. Especially of course my family; the three girls who saw far less of their father than they should have and who sat for hours in a baking hot car outside hospitals as I 'just popped in' to check on patients. Joan who could never rely on my presence at family events, parent-teacher conferences, school sports-days or at meals, but on whom I could always rely for comfort, support and sensible advice.

My long-suffering partner Jake, now gone to his maker. He came from a different background to me; he was a surgeon, a soldier and a conversationalist. I'm afraid that I never really understood him and perhaps that was my loss.

Sue Page, a long term friend and now the editor of this book and my daughter Sascha who created the cover artwork.

Patients in Liverpool, Leeds and Hong Kong who trusted me with their own and their family's health and even lives. They taught me so much about medicine, life and ultimately about myself.

Thank you to all of these.

Chris Howard,
S.W. France. 2022.

GLOSSARY.

Amah. A female domestic servant, usually an elderly Chinese lady with an appallingly bad command of English who speaks in a rapid pidgin. Preferred mode of dress is black trousers and a white tunic, so may be called 'black and white amah'. Often left to raise her employer's children who usually adored her, learnt 'kitchen' Cantonese and acquired a taste for congee. Nowadays amahs are called 'domestic helpers' and come from other places in SE Asia.

Amah's Tooth. Office tower with golden glass cladding which earned its name from a supposed likeness to the single gold stumps that often adorned the mouths of old ladies

Cantonese. The people and language of south China. Guangdong-wah is far less refined than the official language of China, Mandarin or Putong-wah. It has a rich constantly changing argot. They say you can date exactly when a Cantonese speaker left Hong Kong from his use of slang. It sounds dreadful, unlike the smooth tones of Mandarin

Carrian. A group of companies founded by George Tan in 1977 that ended in a scandalous bankruptcy in 1983. Tan managed to convince several large banks that he had wealth to invest and bought several buildings, including Bank of America Tower; businesses; property developers; a bus company and a restaurant. His eventual fall led to a spectacular suicide and many bankruptcies. Several bankers were put in gaol and great interest was taken by all. In actual fact he had no backers and no money, it was a great ride while it lasted.

Catty. Chinese unit of weight used in the wet-markets when you buy fish or veggies. Equivalent to 1.3 lbs or whatever the stall-holder decides.

Central. The main business and smart shopping district on Hong Kong Island. Dominated by bank buildings, government offices, the Hong Kong Club and City Hall on the waterfront.

Clubs. The British, in the days of Empire always had a club where they could drink chota-pegs and eat tiffin out of the sight of the natives. Hong

Kong was no different. They provide not only eating and drinking but sports facilities too. Swimming in the hot months, tennis in the cool.

Colony. Founded in the mid-1800s following the injustice of the Opium Wars under what was regarded by China as an unequal treaty. The term was to become an embarrassment and was replaced by the term Territory. After 1997 it was again changed to Special Administrative Region or SAR.

Congee. A thin rice gruel loved by all Chinese and loathed by practically everyone else. Often made worse (or better if you are Chinese) by flavouring with fish.

Continuity of Care. A system where one knows one's doctor and they know you. Absolutely vital to good medicine and the standard to which all doctors are trained. Under the present NHS (2022) cynically abandoned in the name of efficiency, expediency and Covid.

Dollar. The Hong Kong Dollar was originally fixed at £1 to HK$14. Later it was allowed to float on the markets and rose in value as a result. After a hostile run against it in 1983, it was pegged at HK$7.8 to US$1; a considerable revaluation. The peg has provided financial stability for nearly 40 years.

Elsie Elliot (Elsie Tu). Arrived in Hong Kong as a teacher and missionary in 1951. Elected to the Urban Council, she became a political activist supporting the poorest in society. She was deeply involved in the Star Ferry Riots and a thorn in the side of the British administration all her life. Later the British honoured her. A great woman.

Ergometrine. An injectable drug that causes the uterus to contract. Routinely used after the delivery of a baby to prevent bleeding.

Gweilo. Meaning 'White Ghost' in Cantonese. A slightly insulting term for a white man. Most Europeans embrace the title and ironically use it to describe themselves. Gweipo is the female of the species.

Handover. 1997 was the year the 99-year lease on the NT expired, the rest of the colony, although lacking any limiting date, could not function without this hinterland. Handover in 1997 was inevitable and despite bold attempts to avert some of the worst consequences, the 'One County Two Systems' concept had to be accepted and the British administration left on June 30th, 1997.

Happy Valley. A flat area of land on the north side of the Island. Until the Jockey Club turned it into a race course it was occupied by paddy-fields and was rife with mosquitos. When the first Colonialists built their barracks there, they wondered why so many soldiers died from Malaria. On the slopes of the valley are several cemeteries dating from this time.

ICAC. Independent Commission against Corruption. Set up by Jack Cater in the early 1970s to root out corruption in public life, particularly in the police force. It had sweeping powers of arrest and the seizing of assets. Using the new law that a public servant could not 'be in control of resources exceeding his official emoluments' it removed and imprisoned many senior policemen including Peter Godber then the commissioner of police and also many of his henchmen.

ID card. Everyone over the age of 11 has one. Adults should carry it all the time. You have to show it at the bank, in the post office, at the library. You cannot function without one. A great convenience as it immediately establishes your identity.

IIs, Illegal Immigrants. There is a strong impression that many citizens of the People's Republic were unhappy with the leadership that had brought them such benefits as the Great Leap Forward and the Red Guards. As a result they tried to come and live in Hong Kong 'where rich men eat fat pork'. Most were caught at the border by Ghurkha patrols and repatriated. However, under the Touch Base Policy those that successfully got as far as crossing Boundary Street in Kowloon were allowed to remain and to stop exploitation by Triads they were issued with ID cards.

Jockey Club. A private club devoted to the exploitation of equines. With two race courses, one in Happy Valley the other in the New Territories at Shatin it runs regular race meetings in the cooler months enjoyed by huge crowds. It has a monopoly on local gambling and its huge profits finance equally large charitable functions. Members in return for high monthly fees enjoy luxurious restaurants at the race courses and the Beas

River country club where they can stable their own horses and compete in gymkhanas.

Kai Tak. The famous airport that used to jut out into the Harbour. Planes landing from the North West had to make a steep turn at low altitude to avoid the mountains, then fly between the sky scrapers to land. A well-executed landing often received a cheer from the passengers, if it was botched then with engines on full power the plane would turn for another nail-biting attempt. It was closed after the Handover when the new airport on Lantau opened.

Kowloon or Kowloon-Side. The part of the city north of the Harbour and separated from the New Territories by the steep escarpment which includes the Lion Rock.

Mama-San. Redoubtable mature woman controlling the girls in a bar or brothel.

Mass Transit Railway. The MTR is the underground railway system first started in the mid-1970s.

New Territories or NT. The area between the Chinese border and Kowloon. 'New' because it was leased for 99 years to Britain in 1898; over fifty years after the rest of the colony. Until the 1970s it was mainly country park and agricultural land with a few reservoirs. It has large tracts of delightful open mountainous terrain for walking and camping. Several new towns have sprung up over the years housing several million people.

NGO, Non-governmental organisation. International bodies set up by the U.N. or local organisations established by Hong Kong charities. They usually take on tasks the government seems to be ignoring. Might be financed from taxes or by charities. They also provide an avenue for aspiring grass-roots politicians to get known and to do useful good works.

NHS. National Health Service in the UK. In 1974 less than 30 years old and already a national institution. Probably the best way to deliver health-care even today, provided the politicians finance it properly and don't meddle. I was brought up in it both as consumer and as provider and I loved it.

Pak-pai. Literally white-plate. In the 1970s private cars had white number plates, taxis red and vans black. So, an unregistered illegal taxi was called a pak-pai. In Repulse Bay the local pak-pai driver called himself Six Fingers. He had an extra digit on one hand of which he seemed very proud.

Patten, Chris. UK politician who proved himself a champion of Hong Kong's ordinary people like me in standing up to the Chinese government and their running dogs in the British Foreign Office. Only history will tell if his efforts gained anything in the long run.

The Peak. Once referred to in the local pidgin as 'Topside'. High mountain dominating the island, cool breezes and glorious sea-views keep the very rich comfortable. Before WW2 Chinese were notoriously banned from living on the Peak. The downside is the frequent cloud and fog that enshroud it while at sea-level lesser mortals enjoy the sunshine.

Pokfulam. A residential district at the western end of the Island. Dominated by HK University. It boasts a tiny reservoir and the Jockey Club has a riding school for both able and disabled people. Until the 1960s a large area was cultivated for grass to feed a herd of dairy cows restricted to cattle sheds, the cows later moved to the New Territories.

Queen Mary Hospital. The main teaching hospital situated in Pokfulam. A huge place with very high standards.

Repulse Bay. A large sandy beach on the south side of the Island surrounded by blocks of luxury flats gradually replacing small pre-war apartments and private houses. Thronged in the summer.

Round Table. A men's club for young professionals. Behind the camouflage of charitable acts, the members eat and drink excessively, enjoying innocent fun and camaraderie. Members are call 'tablers'.

SCMP. South China Morning Post. The local daily rag.

Shouson Hill. A residential area on a mound between Repulse Bay and Aberdeen. In the 1970s a lot cheaper than Repulse Bay. A quiet tree-hung road ran round it where children and dogs could wander safely. Now it's built up, busy and expensive.

Snake Head. People smuggler.

Stanley. A village near Repulse Bay with a very popular and extensive market. Low rise housing blocks line the sandy beach. A holiday atmosphere pervades the place. Sailing and windsurfing abound in Tai Tam Bay nearby.

Star-Ferry Riots. In 1966 there was a major civil disturbance occasioned by a fare increase on the cross-harbour Star Ferry. A year later much more serious rioting was fomented by local communists to coincide with the Red Guards movement in the PRC. In the aftermath many expatriates abandoned Hong Kong and there was a major economic recession. These events revealed the dissatisfaction of the populace with their British rulers including police corruption and poor housing and the government subsequently moved to correct both problems.

Stinking Beancurd. Just what it says. Made with rancid oil and deep fried on unhygienic illegal food carts in the street. Malodorous and occasionally a danger to passers-by as the hot oil splashes around when the owner is chased through a crowd by the food hygiene inspectors.

Syntocinon. A drug used as an IV infusion in the labour ward to induce or speed up labour. Also called Pitocin.

Tai-pan. Important man. Head of a large business.

Tai-tai. An important or wealthy woman with a certain attitude.

Triad. A member of a criminal secret society. In fact, just a hoodlum. Triads are deeply embedded in traditional Chinese society and despite concentrated police work persist. They run all the usual rackets.

Star Ferry. Iconic green and white painted ferry boats constantly criss-crossing the Harbour between Central and Kowloon. The airy upper deck costing a bit more than the smelly lower one where you can see the engines and admire the gymnastics of the seamen handling the huge mooring hawsers.

Valium, diazepam. A tranquiliser that was the most prescribed drug in the 1960s. Later it was recognised as addictive and prescribing was discouraged. Some referred to it as Vitamin V as it was so effective against the myriad psychosomatic symptoms of anxiety.

Viva. Viva Voce or oral exam used to assess a candidate's knowledge and his ability to use it under pressure. A very useful tool in weeding out those with only book learning. Seems to have been generally abandoned except for doctors, seafarers and those seeking a driving licence.

Wanchai. District east of Central, in the 1970s notorious for its girlie bars and love hotels. Every so often invaded by American troops on R&R from the Vietnam War. They were often involved in brawls with the squaddies from the British garrison despite club-wielding military policemen from both nations patrolling the streets. Lockhart Road its main thoroughfare.

WAHK. Windsurfing Association of Hong Kong. Founded in 1979 to represent and organise windsurfing. In 1996 Windsurfer Lee Lai-shan was the first Hong Kong athlete yet to win an Olympic gold medal. She was also world champion that year. This opened the government coffers and they then generously supported the sport thereafter

Yacht Club, RHKYC. Where local yachties and rowers go to drink and occasionally get wet. With bases in Causeway Bay (Kellet Island), Middle Island near Repulse Bay and Port Shelter in the NT it organises racing and tuition as well as longer ocean races to Vietnam and the Philippines.